HOMINID ORIGINS

Inquiries Past and Present

Edited by
Kathleen J. Reichs
Davidson College

UNIVERSITY
PRESS OF
AMERICA

University Press of America, Inc.

P.O. Box 19101, Washington, D.C. 20036

Library of Congress Cataloging in Publication Data
Main entry under title:

Hominid origins.

Outgrowth of a symposium organized by the Dept. of Anthropology, Northern Illinois University in 1976.
Bibliography: p.
1. Australopithecines--Addresses, essays, lectures. 2. Fossil man--Addresses, essays, lectures. I. Reichs, Kathleen J. II. Northern Illinois University. Dept. of Anthropology.
GN283.H65 1982 569'.9 82-20161
ISBN 0-8191-2864-3
ISBN 0-8191-2865-1 (pbk.)

Austra lopithecus boisei

Cast of the Clarke and Tobias reconstruction of OH5.

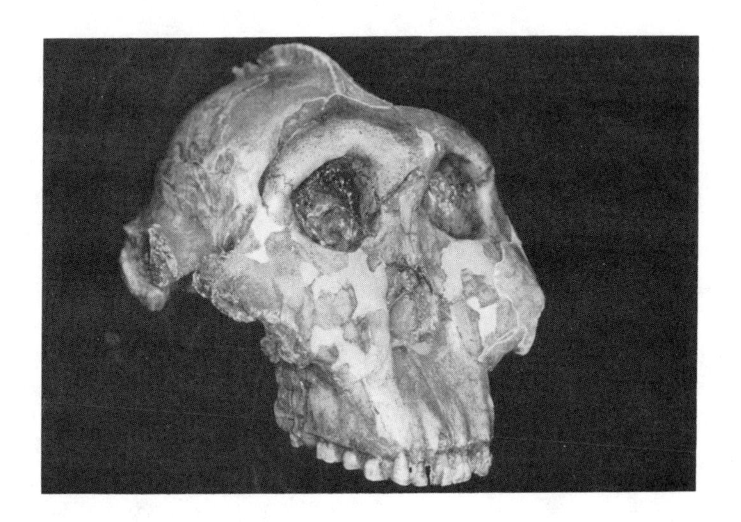

For Kay Toelle

She encouraged me to ask questions.

The author would like to thank Dr. Charlotte Otten who organized the symposium from which these works originated. Her encouragement was the basis for bringing the book together.

TABLE OF CONTENTS

INTRODUCTION

Kathleen J. Reichs
Davidson College

The most unique characteristic of the human species may well be its curiosity, particularly with regard to itself. Human intellects have always sought explanations and answers. Just as the child asks "Where did I come from?", so, too, does mankind question his origins. And, as parents of a less informed age might have squirmed uncomfortably in providing answer for the child's inquiry, likewise did the writings of Charles Darwin effect nineteenth century thinking with an unsettling disquiet. Naturalists such as Buffon, Lyell and Darwin forced a new awareness of man's antiquity. Although the process was slow, the study of human emergence was taken out of the care of the creationists and placed firmly in the hands of the naturalists, where it now resides.

Today the theory of organic evolution is supported by findings in paleontology, embryology, anatomy, physiology and molecular biology. The fossil record, enormously expanded since the time of Darwin, now gives us a fascinating picture of the way we were as we evolved from the small-brained, large-toothed hunter-gatherers of the African savannah to the varied and tremendously successful forms of *Homo sapiens* that people the Earth today.

The study of human evolution has seen a veritable explosion of activity in the last several decades. More data has been collected and analyzed in the past twenty years than in all the preceding period of inquiry into human origins. Not only has there been a rapidly increasing wealth of new fossil meaterials, but whole new areas of investigation have been interwoven into the study of human paleontology.

Geologists have made contributions concerning paleogeography and paleoenvironment, particularly in detailing ancient soil, climate, flora and fauna conditions. Primate studies have helped in the reconstruction of past populations of hominids, with focus at the molecular, anatomical and behavioral levels. Archeology has broadened our understanding of the cultural accomplishments of our earliest ancestors. The study of human origins today is hardly a field devoted to dry library research. The methodologies employed have come to stretch far beyond merely comparing, drawing, measuring and naming fossil specimens, and extend into diverse areas of specialization. The goal has broadened to include reconstruction of both the physiology and behavior of past forms of human life.

One of the most significant new trends in paleoanthropology is the recognition of the importance of ecological factors in the biological evolution and cultural adaptations of man. Here the techniques of geology have been of inestimable help. Geological data concerning fossil sequences and ages, geography, environment, and the processes involved in fossilization such as soil condition, tectonics, deposition and erosion, are critical in reconstructing the paleoenvironmental settings in which hominid populations lived.

New techniques of dating fossil specimens and sites have clarified somewhat the murky picture of interrelationships during critical periods of man's pre-history. The techniques of potassium-argon dating, fission tracking and paleomagnetic stratigraphy increase the reliability of assessing chronology over the more traditional technique of biostratigraphy.

The existence and chronology of environmental oscillations are frequently cited as causal in explaining evolutionary changes. Geological data may best put such hypotheses to the test. Taphonomy (Behrensmeyer 1975), the study of the processes of fossilization, is a relatively new field promising to clarify deposition and erosion sequences, factors so critical in determining specimen and site chronologies. Finally, geology is helping to develop a picture of the lifeways of extinct human groups by providing information about living site locations and faunal associations in relation to regional habitat variability.

Ethological studies currently underway provide an ever-expanding base from which to draw analogy in formulating models of human adaptive patterns in the pre-*Homo* period of evolution. The literature of the last 15 years provides several such models, including those of Lancaster (1967; 1975) based on primate studies, Reynolds (1966; 1968) based on chimpanzee data, Jolly's (1970) seed-eating hypothesis, those of Isaac (1976) and Zihlman (1981) combining primate and archeological information, that of Fox (1967) from the viewpoint of cultural anthropology, as well as several with a strongly biological focus (Washburn 1960; Campbell 1966; Pilbeam

1972a). Each such model deals with factors of bipedalism, tool use, diet, anatomy and behavior as components of an intricately interwoven complex.

Data on non-human primate activities, particularly environmental manipulation and tool use, hunting, infant dependency and familial ties, are particularly relevant. Extending beyond our own order, comparative studies of other life forms, such as those carried out on social carnivores (Kruuk 1972; Schaller 1972; Schaller and Lowther 1969), will even further expand our views of the possibilities and limitations of behavioral patterns in the early stages of human development.

Archeological excavation and analysis is rapidly building a picture of the cultural development and activities of the lower Pleistocene hominids. The recovery and interpretation of artifacts and food refuse, their spatial distribution within sites, and the distribution of sites in relation to local and regional ecology and geography (Isaac 1976:485) is yielding information as to diet, land use, technology and social patterns.

The most significant sites in this time range include those at Olduvai Gorge, Beds I and II (Leakey, M.D. 1971a,b), the Peninj Formation (Isaac 1967), the Shungura Formation (Coppens, Chavaillon and Beden 1973; Chavaillon 1973), the Koobi Fora Formation (Leakey, M.D. 1970a; Isaac et al 1971; Isaac et al. 1971; Isaac et al. 1975), Sterkfontein (Leakey, M.D. 1970b; Mason 1962; Robinson 1962; Tobias and Hughes 1969), Swartkrans (Brain 1970; Leakey, M.D. 1970b), and Makapansgat (Dart 1957a,b; Tobias 1967b). Areas with Pliocene-Pleistocene deposits yielding extraordinary fossil series are being investigated at Kanam and in the Afar depression (Isaac and McCown 1976:427; Johanson, Taieb & Coppers 1982).

Hominid remains have been found consisting of actual living floors, butcher-kill sites, sites composed largely of diffused material, and river and stream channel sites, creating clear evidence of an adaptive pattern involving toolmaking, hunting and food sharing (Isaac 1976:427) as far back as 1.8 million years. There are, however, differences of opinion in interpreting the diversity of stone tool assemblages found in some areas, such as at Olduvai's Bed II. A question which remains unresolved concerns whether differences in stone artifact assemblages reflect different cultural systems showing high degrees of conservatism over long periods of time, or whether they are the result of different activities of a single widespread cultural system.

A second problem as yet unresolved is that of determining the hominid creators of these, the earliest technologies. There is, again, some degree of disagreement on points of interpretation concerning associations between cultural remains and actual hominid forms. Particularly, there are varying opinions as to whether the tools may be attributed to Homo or Australopithecus (Mason 1962; Oakley 1968; Robinson 1962; Tobias 1965). Mary Leakey states that Homo habilis has been found associated with Oldowan tools at no less than five sites in Bed I and one site in the lower portion of Bed II at Olduvai Gorge (Leakey, M.D. 1976:455). Although tools similar to those in Bed II at Olduvai are also found in South Africa at Sterkfontein and Swartkrans, their associations are much less clear.

Another question to be resolved concerns the use of bone as a major resource material in early tool manufacture. While some feel bone was utilized with great frequency (Brain 1967a,b; 1969; 1976; Dart 1957a,b), others are less convinced. Zihlman (1981) stresses the importance which tools of organic composition would have had for early hominids.

Archeology today is not just stone artifact description and classification, but an attempt to understand cultural adaptations and behavioral patterns no longer extant. While the answers are not all in, they are, hopefully, becoming clearer. When coupled with geological data and data concerning closely related species or living species inhabiting similar habitats, anthropologists are coming closer to having a capability to reconstruct a picture of Plio-Pleistocene hominid life. The most direct form of evidence, however, remains the fossils themselves. It is from them that we may best make deductions concerning anatomical, physiological and behavioral adaptations in the course of human evolution.

Perhaps nowhere in anthropology can one find more variability as to interpretations than in the area of Plio-Pleistocene hominid phylogeny. In the earliest phases of hominid development, reaching back into the Miocene, the primary question concerns the point of pongid-hominid divergence and its time depth. While the fossil record suggests this split took place early, in mid-Tertiary times perhaps 20 million years ago, recent developments in molecular biology, including chromosome, serum protein and hemoglobin comparisons, indicate a significantly more recent date (Goodman 1963; Klinger 1963). Immunological data similarly suggest a more recent common heritage for pongids and hominids, with a divergence occurring somewhere in the range of five to ten million years ago. (Sarich 1968; Wilson and Sarich 1969; Sarich and Cronin 1976). While the fossil record is rich from approximately five million years ago on, and significant specimens have been discovered dating to periods prior to fourteen million years ago (i.e.: the dryopithecines of the mid-Tertiary [Andrews 1974; Simons 1963; Washburn 1968]), the critical period between fourteen and five million years ago remains relatively unrepresented.

As described by Clark (1976), the main areas for fossils in Africa include the Western and Gregory Rift areas of East Africa, the Maghred of West Africa, a number of localities in South Africa, and several isolated regions in Egypt, the Sahara, Malawa and Rhodesia (Clark 1976:5), with time depth extending back to the Miocene, or approximately 22 million years.

The period from three to one million years B.P. is particularly rich along the Rift Valley as evidenced at Olduvai Gorge in Tanzania, the area of Lake Turkana (formerly Lake Rudolf) in Kenya, the Omo Valley in southwest Ethiopia, and the Afar Depression where the Rift extends toward the Red Sea. The Rift traces human evolution through the earliest stages at Olduvai, Omo, Lake Turkana and the Afar, through the Middle Pleistocene at Olduvai (Day 1971; Leakey, M.D. 1971a,b), and the Middle and Upper Pleistocene in the Kapthurian Beds (Leakey, M.D. et al. 1969), and the Kibbish Formation at Omo (Leakey, R.E.F., Butzer and Day 1969) where some of the earliest *Homo sapiens* have been found.

But what of the earlier period? Who among the hominids is ancestral to man? What fossil form represents the root stock leading to the Hominidae? Why this gap with only a few tantalizing and fragmentary representatives of a period so critical in human evolution?

Louis Leakey (1962) suggested *Kenyapithecus*, as described at Fort Ternan, as candidate for ancestor to the later Plio-Pleistocene hominids. Pilbeam (1969) and Simons and Pilbeam (1965) place *Kenyapithecus* into one genus with *Ramapithecus*. While this lumped designation is now generally accepted, its hominid status is not.

Controversy still exists as to whether the form represents merely a parallel dental adaptation or a true hominid precursor. While *Ramapithecus* exhibits dental and masticatory similarities to the hominids, with reduction in tooth size and facial length (Simons 1969; Tattersall 1969), the fragmentary nature of the material makes phylogenetic assignment extremely tenuous. Andrews and Walker (1976) conclude, justifiably, that assignment to a family within the hominoidea is not yet warranted.

New fossil evidence is needed dating to the Miocene-Pliocene period. The tooth found in the Baringo Basic deposits, Ngorora Formation (8.5-12 million years B.P.), another from the Lukeino Beds (6.5 million years B.P.) and a probable australopithecine jaw fragment from the southwest end of the Lake Turkana Basin at Lothagam (by faunal association: 5 million years B.P.), indicate that deposits of appropriate time depth exist which may bridge the gap between *Ramapithecus* and *Australopithecus*.

For the next period, the main question revolves around the problem of how many contemporary hominid species/genera coexisted in the late Pliocene and early Pleistocene. If there were more than one species contemporaneously, how should the phylogenetic lines be drawn? When did the genus *Homo* emerge as distinct? What became of other contemporary hominid species?

Most paleoanthropologists now recognize morphological distinctions within the category of Plio-Pleistocene hominids originally referred to as "australopithecine," particularly those based on the gracile-robust dichotomy. What is not clear is the exact number of lineages represented in the fossil record for this time period. One, two and three lineage schemes have been proposed. Furthermore, difficulty in obtaining good dates on some specimens exacerbates the problem of whether the gracile and robust forms are successive or contemporancous populations. This problem also exists in differentiating types within the gracile line.

While both gracile and robust hominids are represented in South Africa, dating remains somewhat obscure. Robust forms, once thought to be younger than their gracile cousins, have now been found at Makapansgat (Aguirre 1970), perhaps the oldest site in South Africa. Taung, once thought to be considerably older, may now turn out, on the basis of geomorphological evidence, to be closer to .87 million years in age (Tobias 1973b), although the validity of this claim has been questioned (Partridge 1973).

New developments in faunal correlation (Freedman and Brain 1972; Maglio 1973; Vrba 1974; Wells 1969) and geomorphological methods (Butzer 1974; Partridge 1973; 1974) may soon untangle the chronology of South African sites.

The story of human paleontology in East Africa began with the discovery of "Zinjanthropus" in 1959 in Bed I of Olduvai Gorge (Tobias 1967a). Shortly thereafter, the fragmented skull of a juvenile, 15 handbones, a clavicle and an almost complete foot were found in apparent association with worked stone tools at the same locality, though slightly below (FLKNN1). (Leakey, Tobias and Napier 1964). The foot belonged, clearly, to a bipedal form; the hand was capable of both power and precision grips (Napier 1962). Thus the specimen was designated *Homo habilis*. The *Homo habilis* designation has since been given to hominid remains from other localities at Olduvai, including DKI (Bed I), FLKII and MNKII (Bed II).

The relationship of *Homo habilis* to other gracile and robust hominid specimens is somewhat problematical. Are they contemporaries or chronologically separate? Research in the East Turkana area and at Omo is helping elucidate the situation.

The Omo Research expedition led by F. C. Howell and Yves Coppens has produced over 100 hominid specimens from the Usno Formation, estimated at between 2.5-2.6 million years B.P. (Brown 1972), and the Shungura Formation, members B through L extending in time from 3.75 to 1.3-1.5 million years B.P. (Most of the hominid specimens have come from the upper members, C through G, with K-Ar dates of approximately 2.5-1.7 million years B.P.). While many of these discoveries are isolated teeth, there are also cranial and mandibular fragments and some post-cranial materials.

Both gracile and robust hominids appear to be represented at Omo, and some specimens have been tentatively categorized as *Australopithecus boisei*, others as either *Australopithecus aff. africanus* and/or *Homo habilis* (Coppens 1970-71; Howell 1968; 1969). Howell and Coppens, among others, (1973; 1976) suggest the existence of more than one hominid lineage at Omo

(Howell 1976; Johanson et al. 1976; Rak and Howell 1978).

Richard Leakey has shown the East Turkana area in northern Kenya to be an extremely valuable site for hominid research. Portions of the stratigraphy overlap the upper sequence at Omo, and faunal evidence suggests an age equivalence with the lower beds at Olduvai. The KBS Tuff, originally dated at 2.6 million years (Curtis et al. 1975), now appears to be closer to 1.8 million years in age (Drake and Curtis 1979). Since 1968 new discoveries have been made on a regular basis, bringing the total to well over 100 hominid specimens (Day and Leakey 1973; Leakey, R.E.F. 1970; 1971; 1972a,b; 1973a, b; Leakey, R.E.F. et al. 1971; Leakey, R.E.F. and Walker 1973; Leakey, R.E.F. and Wood 1973).

As in South Africa, at Omo and at Olduvai, more than one lineage of hominid appears to demonstrate considerable time depth at East Turkana. Robust forms, represented by cranial and mandibular portions and post-cranial materials, have emerged from the lower beds of Koobi Fora and Upper and Illeret members. Gracile forms (*Homo*?), showing greater morphological variability, especially if viewed in relation to the chronology of the beds, (early to late) (Clark 1976) also inhabited the area for long periods of time.

One of the most spectacular of the finds at East Turkana, was Richard Leakey's KNMER-1470 cranium (Leakey, R.E.F. 1973b) from the lower Koobi Fora Formation. The specimen consists of an almost complete cranium with much of the face and palate, though lacking teeth. Surprinsingly, its endocranial capacity falls close to 800 cc, double the mean for the South African *A. africanus* forms (Holloway 1970; 1972b; 1973; Robinson 1966; Tobias 1971), and one quarter again as large as *Homo habilis* at Olduvai (OH4) (Holloway 1973; Tobias 1971). (Others have placed the capacity closer to 750 cc [Brace et al. 1979]). KNMER-1470 derives from below the KBS tuff dated to 1.8 million years B.P., and, therefore, appears roughly contemporaneous with the Olduvai representatives of *Homo*.

Also present at East Turkana are representatives of what Richard Leakey (1973b) views as a possible third hominid lineage. KNMER-1813, discovered in 1973, is the cranium of a smaller-brained (500 cc) gracile hominid, probably from the lower member of the Koobi Fora Formation. If he is correct in suggesting that this form shows resemblance to *Australopithecus africanus*, then three hominid lineages existed simultaneously in the early Pleistocene at East Turkana. In any event, both the gracile and robust lineages seem to have coexisted in the East Turkana area for a considerable time depth.

Other evidence has emerged of early large brained (*Homo*?) hominids by the lower Pleistocene. A large skull from the Chad basin (Coppens 1967) and fragmentary cranial and mandibular portions from Swartkrans (Clark 1976; Clark and Howell 1972; Clark et al. 1970) both support the possibility of the contemporaneity of an advanced hominid along with the robust hominid forms.

In the mid-1970's news spread of a new hominid local in the Afar triangle area of northeastern Ethiopia. In 1974 "Lucy" was introduced to the world as the oldest and most complete hominid skeleton to date, followed the next year by the so-called "first family" (Johanson, D. 1976; Johanson, D. and M. Taieb 1976). The International Afar Research Expedition has subsequently published over 80 works describing the 240+ hominid specimens from this area. (For a complete chronological listing see the April 1982 edition of the *American Journal of Physical Anthropology*.) A minimum age of over 3 million years has been assigned to these materials. The same year Mary Leakey announced similarly dated fossils (3.5-3.7 million B.P.) at Laetoli (Leakey et al. 1976), and in 1979 added information concerning footprints at the same site--the oldest evidence of hominid bipedalism.

Interpretations of the Hadar-Laetoli hominid series vary. Johanson and White (1978; 1979), primarily on the basis of dental features, place all of the material into a new taxon, *Australopithecus afarensis*, and view this as the sole lineage for the time range 3-4 million B.P. period. Mary and Richard Leakey disagree, stating that the Laetoli materials show greater similarity to early *Homo* specimens of East Africa. (Leakey et al. 1976). They see at least two contemporaneous hominid forms for this time period: *Australopithecus* and *Homo*.

Clearly, taxonomy and phylogenetic relationships among the late Pliocene-early Pleistocene hominids are anything but clear. While most agree that the robust (*Australopithecus robustus/boisei*) forms represent a distinct lineage, there remains considerable controversy over the significance of morphological variability among the gracile forms (Campbell 1972; Pilbeam 1972a; Robinson 1967; 1972; Tobias 1968; 1973b; 1976). Most schemes suggest the presence of two genera of late Pliocene and early Pleistocene hominids: *Australopithecus* and *Homo*. Tobias (1973a) suggests one genus, *Australopithecus*, composed of one superspecies, *Australopithecus robustus* and *Australopithecus boisei*, and one polytypic species, *Australopithecus africanus*. Others who either regard the gracile and robust forms as members of one genus, *Australopithecus*, or who recognize only two hominid genera, include Campbell (1972), Pilbeam (1972a,b), Simons (1972), Simons and Ettel (1970), Simpson (1963) and von Koenigswald (1973).

Robinson (1972) continues to make distinction at the generic level between *Homo* and *Paranthropus*, based on proposed differences in adaptive patterning. Others argue that such distinctions are unjustified (Campbell 1972; Mayr 1963; Tobias 1967). Several workers point to simple body size differences and the phenomenon of allometry; or differential growth, as the cause of dental differences in gracile and robust hominids (Brace 1972; McHenry 1975; Pilbeam and Gould 1974; Wolpoff 1974). Others disagree, maintaining that the teeth most likely indicate dietary differences (DuBrul 1977; Wood and Stack 1980).

Clark (1976) points out that while the robust hominid populations appear to remain morphologically unchanged throughout their existence, representative specimens of the gracile lineage demomstrate marked change over time. The Bed II *Homo habilis* specimen from Olduvai Gorge, for example, is more advanced (more similar to *Homo erectus* as at LLK II in Bed IV) than those derived from Bed I. In contrast the robust lineage appears quite stable and, in fact, disappears by upper Bed II (Clark 1976). His conclusion, that the period in question was one of extremely rapid evolution for the gracile line, serves as a caveat to those attempting taxonomic assignment of fossil specimens. Similarly, the recent recognition of the great variability possible in hominid populations must of necessity dictate careful consideration before phylogenetic interpretations can be made.

Furthermore, until accurate assignment of dates is possible, reconstruction of hominid relationships must remain tenuous at best. If, for example, the South African fossil materials turn out to be contemporaneous with Bed I at Olduvai or later, then the *Australopithecus africanus* forms could not be ancestral to the *Homo habilis* forms, but might represent a mutually exclusive and contemporaneous third lineage.

The issue is far from settled. There remain proponents of the one lineage theory (LeGros Clark 1967; Wolpoff 1971), the two lineage theory (Campbell 1972; Pilbeam 1972a,b; Simons 1972), and the three lineage theory (Clark 1976; Leakey, R.E.F. 1976), each group placing differing degrees of emphasis on the morphological variability exhibited in the Plio-Pleistocene hominid fossil record.

The task of reconstructing the course of human evolution is an ongoing one. Each day new fossil discoveries are being made. The areas of Hadar in Ethiopia (Johanson and Taieb 1976; Johanson, White and Coppens 1978; Johanson, Taieb and Coppens 1982), Laetoli in Tanzania (Leakey, M.D. et al. 1976; White 1977), and Chesowanja in Kenya (Carney et al. 1971) are good examples of promising new locales for Plio-Pleistocene human paleontology.

Miocene-Pliocene deposits are also being discovered in Turkey (Andrews and Tabien 1977), Saudi Arabia (Andrews et al. 1978), Pakistan (Pilbeam et al. 1977) and Eastern Europe (deBonis and Melentis 1977; Kretozoi 1975), which may help fill the critical gap spanning 15 to 5 million years B.P.

Extensive examination of new hominid materials, as well as re-examination of earlier discoveries, is helping to assemble a more accurate picture of the complete morphological pattern and capabilities of early hominids. The question of bipedalism, its earliest appearance as a hominid locomotor adaptation, the range of variability present during the Plio-Pleistocene period of development and the degree of difference between the morphological adaptation of these groups and Homo sapiens, is being investigated (Charteris, Wall and Nottrodt 1982; Johanson and White 1979; Johnson et al. 1978; Leakey, M.D. 1978; Leakey, M.D. and Hay 1979; Lovejoy 1978; McHenry and Corruccini 1976; 1978; McHenry and Temerin 1979; Prost 1980; Robinson 1978; Thompkins 1977; Wolpoff 1976; Zihlman and Brunker 1979). Several workers are attempting to calculate hominid stature (Olivier 1976; Reed and Falk 1977; Steudel 1980). The functional capability of the Pleistocene hominid hand is being reassessed (Bush 1980; Sussman and Creel 1979). Examination continues of the foot (Oxnard and Lisowski 1980), upper and lower limb proportions (McHenry 1978), the upper arm (Feldesman 1979; McHenry 1976, the scapula (Vallois 1977; Vrba 1979), shoulder (Ciochon and Corruccini 1976), sacrum (Leutenegger 1977), teeth

(Wolpoff 1979; Wood and Stock 1980), and brain (Deacon et al. 1980; Falk 1979; 1980), to name but a few. All will result, ultimately, in a more complete appreciation of the functional make-up of early man. And, of course, the phylogenetic relationships of the newest finds are being deciphered (Kimball 1980; McHenry and Corruccini 1980; Pilbeam 1980). Finally, ongoing geological studies (Aronson et al. 1977; Ayala 1976; Brown 1978; Hay 1976) will place this information on firmer chronological and ecological footing.

Geology, ethology, archeology and paleontology, together, have given us the picture we now hold of the earliest phases of human evolution. The fossils themselves, the artifacts they have left behind, the remains of their meals, and the features of the sites they created are the main data banks from which we draw. Evidence concerning environmental and climatic conditions are equally essential. And, finally, information on the behavioral systems of living species and peoples assist in completing the picture of an adaptive complex long vanished.

This volume, then, is intended to present an overview of this multi-faceted approach. It includes articles on the early work done with australopithecine material, as well as interpretations based on the most recent finds. It is the outgrowth of a symposium organized by the Department of Anthropology, Northern Illinois University in the spring of 1976. While the main theme of the conference was "australopithecine" research, the participants came from diverse backgrounds. There, as here, the aim was to demonstrate the wide diversity of specializations which come to bear on the questions of human origins. Each article, while attempting to add to a better appreciation of the Plio-Pleistocene hominid adaptation, approaches the topic from a different point of view. The methodologies are diverse, the interpretations varied, but the goal universal.

The articles by Reed and Sigmon focus on historical aspects of hominid research. Butzer deals briefly with geological factors. Wolpoff and Protsch discuss the significance of functional and morphological characteristics in determination of phylogenetic relationships. In the final two chapters, Oxnard and Zihlman focus on the use of comparative primate data in the interpretation of hominid adaptation, the former with regard to morphology, the latter concerning behavior.

In the opening article Charles Reed provides a detailed overview of the earliest phases of research concerning forms which are or were at one time referred to as Australopithecus, Pleisianthropus or Paranthropus. He focuses on the period 1924-1951, reflected mostly in the works of Raymond Dart and Robert Broom, and discusses their struggle to overcome the rejection of "australopithecus" as a member of the Hominidae. His detailed historical recreation sheds a great deal of light on the sources and causes of the high degree of taxonomic confusion which typified this era. Reed points out, for example, the unfortunate consequences of Dart's use of the term "ape" to refer to the Taung specimen. Not only did this cast a shadow over the fossil with regard to taxonomic assignment, but the ill-defined use of the labels "man-ape" and "ape-man" persisted long in use, further confusing the situation.

Reed provides insight into the historical and philosophical background which helped shape early taxonomic interpretations, many of which, to some extent, linger today. His bibliography is invaluable to anyone interested in analyzing early controversies concerning the australopithecines and their status as human ancestors.

Becky Sigmon's article centers on a later point in time and shifts the focus from South Africa to East Africa. She outlines the problems now facing paleoanthropology, particularly with regard to the relationships of the fossil populations from these two geographic areas. She discusses the current emphasis on multi-disciplinary studies and the resulting abundance of new data, including more fossils, better absolute dates and more accurate environmental information.

Sigmon discusses how the recent increase in recovered fossil specimens has created additional taxonomic concerns, and elaborates on designations such as "Zinjanthropus," "Homo habilis" and "Homo ergaster." She considers the question of the co-existence of two or three sympatric hominid lineages, and the differences of opinion with regard to bipedalism and cultural development.

While Reed's article deals mainly with the phase of acceptance, Sigmon's deals with the next phase of human paleontology, that of discovery. She concludes by suggesting that the next phase, upon which we are now entering, is one of interpretation and an increasing understanding of hominid environmental interaction.

In his contribution Karl Butzer, in keeping with the multi-disciplinary approach, presents evidence concerning the environmental conditions prevalent in South Africa during the late Pliocene and early Pleistocene. In comparing geomorphological data at Taung, Makapansgat, Sterkfontein, Swartkrans and Kromdrai, he concludes that australopithecine populations inhabited open, grassy environments interspersed with bush, thickets or woodlands. He states that there is no geological evidence supporting an hypothesis of dietary differences between gracile and robust hominid forms, and presents a model of contemporaneous, sympatric hominid lineages.

Milford Wolpoff's article centers on the functional interpretation of the hominid fossils themselves. He draws primarily from evidence uncovered in the past 10 years in arguing for a phyletic interpretation of australopithecines as direct human ancestors. He questions those who would exclude the australopithecines based on multivariate analyses, and cautions against equating morphometrics with function. He similarly disagrees with those who would exclude australopithecus from ancestry based on belief in a more advanced Plio-Pleistocene hominid form. Wolpoff points to morphological evidence to support his view of an australopithecine stage of human evolution.

R. R. Protsch's contribution deals, also, with morphological interpretation and the consequent implications for taxonomic assignment, although the specimens he is discussing are hardly recent discoveries. In describing the Eyasi and Garusi finds from East Africa he questions their earlier designations as members of Homo erectus and Meganthropus africanus respectively. Based primarily on dental morphology and development, he suggests affinities for the latter to the Pre-Homo erectus populations of East Africa. Similarities are particularly striking, he points out, when compared to the Laetolil material recently discovered by Mary Leakey.

Protsch's approach demonstrates the necessity for re-evaluation of old materials and ideas. His reassessment of the Garusi and Eyasi specimens typifies the kind of self-critical attitude so necessary in the formulation, testing and restructuring of scientific hypotheses. His description of the cleaning, dating, reconstructing and use of x-ray analysis on the materials in question shows how new techniques can be applied to old materials, with extremely fruitful results.

The use of functional relationships of living species for the interpretation of fossil anatomical structure is presented in the section by Charles Oxnard. He demonstrates the use of biometrics and multi-variate statistical analysis in comparing forelimb and hindlimb structure and overall body proportions. His findings indicate significant correlation between multi-variate studies of primate measurements and general primate taxonomy as derived from morphological, physiological and biomolecular observations.

In discussing the implications of this approach for hominid research, Oxnard draws conclusions concerning australopithecine patterns of locomotion. Basing his view on analysis of the pelvis, talus and clavicle, he proposes a type of locomotor capability for early hominids that was quite unique, one combining climbing and walking adaptations. Such a combination is seen in no other primate. He suggests that this upright but climbing hominid would have best fitted the environment from which his fossil remains appear to derive. He suggests the possibility of relegating the australopithecines to a side role in human evolution, and removing them from a position of direct ancestry to *Homo*. Oxnard concludes by calling for a new assessment of hominid data, emphasizing the need for total data recognition and an overall approach to total form in fossil analysis.

In the final article Adrienne Zihlman presents a behavioral reconstruction of Plio-Pleistocene hominid groups. She summarizes the uniquely human elements of adaptation as bipedalism and tool use, along with a long childhood and increased intelligence, and relates these to the "australopithecine" level of development. In formulating a model of australopithecine subsistence activities she emphasizes the importance of gathering, as opposed to hunting, a deviation from the more traditional view of early human activities. According to her model the earliest adaptive pattern would have been one of foraging, with hunting activities coming later.

It was the eventual development, in a savannah environment, of an omnivorous diet, based on collecting and sharing, which was critical in shaping social behavior at this stage. By using data derived from observations of non-human primate species and contemporary foraging human groups, Zihlman extrapolates from fossil anatomy to behavioral reconstruction.

This volume, then, contains a representative sample of hominid research strategies. It is from studies such as these that we get our picture of early hominids as large toothed, small brained, bipedal creatures who used tools, gathered and hunted, established home bases, shared, and took great pains to raise their young. The diversity exhibited here reflects the diversity of approaches in human paleontology, past and present. In the varying interpretations of the authors one can gain an appreciation for the fascinating possibilities in the interpretation of human evolution. While we have come a long way since the discovery of the child at Taung in 1924, we have yet a long way to go.

Literature Cited

Aguirre, E. 1970. Identification de "Paranthropus" en Makapansgat Cronica del XI Congreso de Arqueologia, Merida. 1969, pp. 98-124.

Andrews, P. 1974. New species of Dryopithecus from Kenya. Nature 249:118-190.

Andrews, P., W. R. Hamilton & P. J. Whybrow. 1978. Dryopithecines from the Miocene of Saudi Arabia. Nature 274:249-251.

Andrews, P. and N. Tobien. 1977. New Miocene locality in Turkey with evidence on the origin of Ramapithecus and Sivapithecus. Nature 268:699-701.

Andrews, P. and A. Walker. 1976. The primate and other fauna from Fort Tennan, Kenya. In: Human Origins: Louis Leakey and the East Africa Evidence, G.Ll. Isaac & E. McCown, eds., pp. 279-304; W. A. Benjamin, Inc.

Aronson, J. L., T. J. Schmitt, R. C. Walter, M. Taieb, J. J. Tiercelin, D. C. Johanson, C. W. Waeser and A. E. M. Nairn. 1977. New geographical and paleontological data for the hominid-bearing Hadar Formation, Ethiopia. Nature 267:323-327.

Ayala, F. J. 1976. Molecular Evolution. Sinaeur Associates. Sunderland, Massachusetts.

Behrensmeyer, A. K. 1975. The taphonomy and paleoecology of Plio-Pleistocine vertebrate assemblages East of Lake Rudolf, Kenya. Bulletin of the Museum of Comparative Zoology 146:473-578.

Brace, C. L. 1972. Sexual dimorphism in human evolution. Yearbook in Physical Anthropology.

Brace, C. L., H. Nelson, H. Korn & M. L. Brace. 1979. Atlas of Human Evolution, Holt, Rinehart & Winston.

Brain, C. K. 1967a. Bone weathering and the problem of bone pseudo-tools. S. Afr. J. Sci. 63:97-99.

Brain, C. K. 1967b. Hottentot food remains and their bearing on the interpretation of fossil bone assemblages. Scientific papers of Namib Desert Research Station, Pretoria, South Africa 32:1-7.

Brain, C. K. 1969. The contribution of Namib Desert Hottentots to an understanding of Australopithecine bone accumulations. Scientific papers of Namib Desert Research Station, Pretoria, South Africa 39:13-22.

Brain, C. K. 1970. New finds at the Swartkaans Australopithecine Site. Nature 225:1112-1119.

Brain, C. K. 1976. Some principles in the interpretation of bone accumulations associated with man. In: Human Origins: Louis Leakey and the East African Evidence, G. Isaac and E. McCown, eds., 97-116, W. A. Benjamin, Inc.

Brown, F. H. 1972. Radiometric dating of sedimentary formations in the Lower Omo Valley, Southern Ethiopia. In: Calibration of Hominoid Evolution. W. W. Bishop & J. A. Miller, eds., Edinburgh: Scottish Academic Press, 273-287.

Brown, F., F. C. Howell & G. G. Eck. 1978. Observations and problems of correlation of Late Cenozoic hominid bearing formations in North Lake Turkana Basin. In: Geographical Background to Fossil Man. W. W. Bishop, ed., Toronto: University of Toronto Press, pp. 473-498.

Bush, M. E. 1980. The thumb of Australopithecus Afarensis. Paper presented at the Forty-ninth Annual Meeting of the American Association of Physical Anthropology, Niagara Falls.

Butzer, K. W. 1974. Paleo-ecology of South African Australapithecines: Taung Revisited. Current Anthropology.

Campbell, B. G. 1966. Human Evolution: An Introduction to Man's Adaptations, Chicago: Aldine.

Campbell, B. G. 1972. Conceptual progress in physical anthropology: fossil man. Ann. Rev. Anthropo. 1. 1:27-54.

Carney, J., A. Hill, J. A. Miller & A. Walker. 1971. Late Australopithecine from Baringa District, Kenya. Nature 230:509-514.

Charteris, J., J. C. Wall, J. W. Nottrodt. 1982. Pliocene hominid gait: New interpretations based on available footprint data from Laetoli. Am. J. Phys. Anthrop. 58:133-144.

Chavaillon, J. 1975. Evidence for the technical practices of early Pleistocene hominids: Shungura Formation, Valley of the Omo, Ethiopia. In Earliest Man and Environments in the Lake Rudolf Basin. Y Coppens, F. C. Howell, G. Ll. Isaac, and R.E.F. Leakey, eds. Chicago: University of Chicago Press.

Ciochon, R. and R. S. Corruccini. 1976. Shoulder joint of Sterkfontein Australopithecus. S. Afr. J. Sci. 72:80-82.

Clark, J. D. 1976. The African origins of man the toolmaker. In: Human Origins: Louis Leakey and the East African Evidence, G. Ll. Isaac & E. McCown, eds., pp. 1-53. W. A. Benjamin, Inc.

Clark, R. J., F. C. Howell & C. K. Brain. 1970. More evidence of an advanced hominid at Swartkruns. Nature 225:1219-1222.

Clark. R. J. and F. C. Howell. 1972. Affinities of the Swartkrans 847 hominid Chanium. Am. J. Phys. Anthrop. 37:319-336.

Clark, LeGros W. E. 1967. Man-Apes or Ape-Men? New York: Holt, Rinehart & Winston, Inc.

Coppens, Y. 1967. L'hominien du Tchad. Actal del v Congreso Panafricano de Prehistoric y de Eustudio del Cuaternario. L. D. Cuscoy, ed., Tenerife, 1963. pp. 329-330.

Coppens, Y. 1970-71. Localisation dans le temps and dans l'espace des restes d'hominides des formations Plio-Pleistocenes de l'Omo (Ethiopie). C. R. Acad. Sci., Paris 271:1968-71, 2286-2289; 272:36-39.

Coppens, Y., J. Chavaillon & M. Beden. 1973. Resultats de la nouvelle Mission de l'Omo (campagne 1972). Decouverte de restes d'hominides et dune industrie sur eclats. Comptes rendus de l'Academie des Sciences, Paris (Serie D) 276:161-164.

Curtis, G., R. Drake, T. Cerling & A. Hampel. 1975. Age of KBS tuff in Koobi Fora Formation, East Rudolf, Kenya. Nature 258:395-398.

Deacon, T. W., A. G. Filler & J. E. Cronin. 1980. The rate and nature of brain evolution in the Hominidae. Paper presented at the Forth-ninth Annual Meeting of the American Association of Physical Anthropology, Niagara Falls.

Dart, R. 1957a. The Makapansgat Australopithecine Osteodontokeratic culture. In: Proceedings of the Third Pan-African Congress of Prehistory, Livingstone, 1955. J. D. Clark & S. Cole, eds. London: Chatto and Windus. pp. 161-171.

Dart, R. 1957b. The Osteodontokeratic culture of Australopithecus prometheus. Transvaal Museum Mem. 10:1-105.

Day, M. H. 1971. Post-cranial remains of Homo erectus from Bed IV, Olduvai Gorge, Tanzania. Nature 232: 383-387.

Day, M. H. and R.E.F. Leakey. 1973. New evidence of the genus Nomo from East Rudolf, Kenya. I. Am. J. Phys. Anthrop. 39:341-354.

deBonis, L. and J. Melentis. 1977. Un nouveau genre de primate homonoide dans Le Vallesian (Miocene Superieur) de Macedoina. C. R. Aca. Sc. Paris 284:1393-1396.

Drake, R. and F. Curtis. 1979. Radioisotope dating of the Laetoli beds, the Hadar formation and the Koobi-Fora Shungura formations. Paper presented at the Forth-eighth Annual Meeting of the American Association of Physical Anthropology, San Francisco.

DuBrul, E. 1977. Early hominid feeding mechanisms. Amer. J. Phys. Anth. 47:305-321.

Falk, D. 1979. On a new Australopithecine partial endocast. Am. J. Phys. Anthrop. 50:611-614.

Falk, D. 1980. Language, handedness and primate brains: did the Australopithecines sign? Amer. Anthrop. 82:72-78.

Feldesman, M. 1979. Further morphometric studies of the ulna from the Omo Basin, Ethiopia. Am. J. Phys. Anthrop. 51:409-415.

Fox, R. 1967. In the beginning: aspects of hominid behavioral evolution. Man 2:415-433.

Freedman, L. and C. K. Brain. 1972. Fossil Cercopithecoid remains from the Kromdrai Australopithecine Site (Mammalia: Primates). Ann. Transvall Museum 28:1-16.

Goodman, M. 1963. Man's place in the phylogeny of the primates as reflected in serum proteins. In: Classification and Human Evolution. S. L. Washburn, ed. Chicago: Aldine. pp. 204-234.

Holloway, R. L. 1970. Australopithecine endocast (Taung Specimen 1924): A new volumn determination. Science 168:966-968.

Holloway, R. L. 1972. Australopithecine endocasts, brain evolution in the hominidea, and a model of hominid evolution. In: The Functional and Evolutionary Biology of Primates. R. Tuttle, ed. Chicago/New York: Aldine-Atherton, pp. 185-203.

Holloway, R. L. 1973. Endocranial volumns of early African hominids and the role of the brain in human mosaic evolution. J. Human Evol. 2:449-459.

Howell, F. C. 1968. Omo research expedition. Nature 216:567-572.

Howell, F. C. 1969. Remains of hominidae from Pliocene/Pleistocene formations in the Lower Omo Basin, Ethiopia. Nature 223:1234-1239.

Howell, F. C. 1976. An overview of the Pliocene and earlier Pleistocene of the Lower Omo Basin, Southern Ethiopia. In: Human Origins: Louis Leakey and the East African Evidence. G. Ll. Isaac and E. McCown, eds., pp. 227-268, W. A. Benjamin, Inc.

Howell, F. C. and Y. Coppens. 1973. Deciduous teeth of hominidae from the Pliocene/Pleistocene of the Lower Omo Basin. Ethiopia. J. Hum. Evol. 2: 461-472.

Howell, F. C. and Y. Coppens. 1976. An overview of hominidae from the Omo Succession, Ethiopia. In: Earliest Man and Environments in the Lake Rudolf Basin. Y. Coppens, F. C. Howell, G. Ll. Isaac and R.E.F. Leakey, eds. Chicago: University of Chicago Press, pp. 522-532.

Isaac, G. Ll. 1967. The stratigraphy of the Peninj group--early middle Pleistocene formations west of Lake Natron, Tanzania. In: Background to Human Evolution. W.W. Bishop and J.D. Clark, eds. Chicago: University of Chicago Press, pp. 229-257.

Isaac, G. Ll. 1976. The activities of early African hominids: a review of archeological evidence from the time span two and a half to one million years ago. In: Human Origins: Louis Leakey and the East African Evidence, G. Ll. Isaac and E. McCown, eds., pp. 483-514, W.A. Benjamin, Inc.

Isaac, G. Ll., R.E.F. Leakey and A.L. Behrensmeyer. 1971. Archeological traces of early hominid activities east of Lake Rudolf, Kenya. Science 173:1129-1134.

Isaac, G. Ll., J.W.K. Harris and D. Crader. 1975. Archeological evidence from the Koobi Fora Formation. In: Earliest Man and Environments in the Lake Rudolf Basin: Stratigraphy, Paleoecology and Evolution. Y. Coppens, F.C. Howell, G.Ll. Isaac and R.E.F. Leakey, eds., Chicago: University of Chicago Press.

Isaac, G.Ll. and E. McCown. 1976. Human Origins: Louis Leakey and the East African Evidence. W.A. Benjamin, Inc.

Johanson, D.C.Y. Coppens, & M. Taieb. 1976. Pliocene hominid remains from Hadar, Central Afar, Ethiopia. Ixe Congres, Untion Int. des Sci. Prehist. et Protohist. Nice, pp. 120-133.

Johanson, D.C. & M. Taieb. 1976. Plio-Pleistocene hominid discoveries in Hadar, Ethiopia. Nature 260:293-297.

Johanson, D.C.,M.Taieb, Yves Coppens.1982. Pliocene Hominids from the Hadar Formation,Ethiopia (1973-77): Stratigraphic, Chronologic & Paleoenvironmental Contexts, with Notes on Hominid Morphology and Systematics; Am. J. Phys. Anthrop., 57: 373-402

Johanson, D.C., T.D. White. 1979. A systematic assessment of early African hominids. Science 202: 321-330.

Johanson,D.C.,T.D. White, Y. Coppens. 1978. A new species of genus Australopithecus (Primates: Hominidae) from the Pliocene of Eastern Africa. Kirtlandia 28: 1-14.

Jolly, C. 1970. The seed-eaters: a new model of hominid differentiation based on a baboon analogy. Man 5:5-26.

Kimbell, N. H. 1980. A reconstruction of the adult cranium of Australopithecus afarensis. Paper presented at the Forty-ninth Annual Meeting of the American Association of Physical Anthropology, Niagara Falls.

Klinger, H. P., J. L. Hamerton, D. Murton and E. M. Lang. 1963. The chromosomes of the hominoidea. In: Classification and Human Evolution. S. L. Washburn, ed., Chicago: Aldine, pp. 235-242.

Kretozoi, M. 1975. New Ramapithecines and Pliopithecus from the Lower Pliocene of Rudabanya in Northeastern Hungary. Nature 257:578-581.

Kruuk, H. 1972. The Spotted Hyena: A Study of Predation and Social Behavior. Chicago: University of Chicago Press.

Lancaster, J. 1967. The evolution of tool using behavior. American Anthropology 70:56-66.

Lancaster, J. 1975. Primate Behavior and the Emergence of Human Culture. Holt, Rinehart & Winston.

Leakey, L.S.B. 1962. A new lower Pliocene fossil from Kenya. Annals and Magazine of Natural History 13: 689-696.

Leakey, L.S.B., P. V. Tobias & J. R. Napier. 1964. A new species of the genus Homo from the Olduvai Gorge. Nature 202:7-9.

Leakey, M.D. 1970a. Early artifacts from the Koobi Fora Area. Nature 226:228-230.

Leakey, M.D. 1970b. Stone artefacts from Swartkrans. Nature 225:1222-1225.

Leakey, M.D. 1971a. Discovery of postcranial remains of Homo erectus and associated artefacts in Bed IV at Olduvai Gorge, Tanzania. Nature 232:380-383.

Leakey, M.D. 1971b. Olduvai Gorge Volumn III. Excavations in Beds I and II, 1960-1963. Cambridge: The University Press.

Leakey, M. D. 1976. A summary and discussion of the archeological evidence from Bed I and Bed II, Olduvai Gorge, Tanzania. In: Human Origins: Louis Leakey and the East African Evidence, G. Ll. Isaac and E. McCown, eds., pp. 431-459, W. A. Benjamin, Inc.

Leakey, M. D. 1978. Pliocene footprints at Laetolil, Northern Tanzania. Antiquity 52:133.

Leakey, M. D. and R. L. Hay. 1979. Pliocene footprints in the Laetolil Beds at Laetoli, Northern Tanzania. Nature 278:317-323.

Leakey, M. D., R. L. Hay, G. H. Curtis, R. E. Drake, M. K. Jackes and T. D. White. 1976. Fossil hominids from the Laetolil Beds. Nature 62:460-466.

Leakey, M. D., P. V. Tobias, J. E. Martyn and R.E.F. Leakey. 1969. An Acheulean industry and hominid mandible, Lake Baringo, Kenya. Proceedings of the Prehistoric Society 35:48-76.

Leakey, R.E.F. 1970. Fauna and artefacts from a new Plio/Pleistocene locality near Lake Rudolf in Kenya. Nature 226:223-224.

Leakey, R.E.F. 1971. Further evidence of Lower Pleistocene hominids from East Rudolf, North Kenya. Nature 231:241-245.

Leakey, R.E.F. 1972a. Further evidence of Lower Pleistocene hominids from East Rudolf, North Kenya, 1971. Nature 237:264-269.

Leakey, R.E.F. 1972b. New fossil evidence for the evolution of man. Social Biology 19:99-114.

Leakey, R.E.F. 1973a. Further evidence of Lower Pleistocene hominids from East Rudolf, North Kenya, 1972. Nature 242:170-173.

Leakey, R.E.F. 1973b. Evidence for an advanced Plio-Pleistocene hominid from East Rudolf, Kenya. Nature 242:447-450.

Leakey, R.E.F., K. W. Butzer and M. H. Day. 1969. Early homo sapiens remains from the Omo River Region of Southwest Ethiopia. Nature 222:1132-1138.

Leakey, R.E.F., J. M. Mungai and A. C. Walker. 1971. New Australapithecines from East Rudolf, Kenya (11). *American Journal of Physical Anthropology* 35:175-186.

Leakey, R.E.F. and A. C. Walker. 1973. New Australopithecines from East Rudolf, Kenya (II). *American Journal of Physical Anthropology* 39:355-368.

Leutenegger, W. 1977. A functional interpretation of the sacrum of *Australopithecus africanus*. *South African Journal of Science* 73:308-310.

Lovejoy, O. 1978. A biomechanical review of the locomotor diversity of early hominids. In: *Early Hominids of Africa*, C. J. Jolly, ed., Duckworth, London, pp. 403-429.

Maglio, G. 1973. Origin and evolution of the elephantidae. *Transcripts of American Philosophical Society*, n.s. 63:3:1-149.

Mason, R. K. 1962. Australopithecines and artefacts at Sterkfontein, Part II. *South African Archaeology Bulletin* 17:109-125.

Mayr, E. 1963. The taxonomic evaluation of fossil hominids. In: *Classification and Human Evolution*, S. L. Washburn, ed., Chicago: Aldine, pp. 332-346.

Merrick, H. V., J. de Heinzelin, P. Haesaerts and F. C. Howell. 1973. Archeological sites of Early Pleistocene Age from the Shungura Formation, Lower Omo Valley, Ethiopia. *Nature* 242:572-575.

McHenry, H. M. 1975. Fossil hominid body weight and brain size. *Nature* 254:686-688.

McHenry, H. M. 1976. Multivariate analysis of early hominid humeri. In: *The Measures of Man*, E. Giles & J. S. Friedlander, eds., Peabody Museum Press, Harvard University, Cambridge, pp. 338-371.

McHenry, H. M. 1978. Fore-and hindlimb proportions in Plio-Pleistocene hominids. *American Journal of Physical Anthropology* 49:15-22.

McHenry, H. M. and R. S. Corruccini. 1976. Fossil hominid femora and the evolution of walking. *Nature* 259:657-568.

McHenry, H. M. and R. S. Corruccini. 1978. The femur in early human evolution. American Journal of Physical Anthropology 49:473-488.

McHenry, H. M. and R. S. Corruccini. 1980. Mio-Pliocene hominoids: Phenetic and Cladistic affinities. Paper presented at the Forty-ninth Meeting of the American Association of Physical Anthropology, Niagara Falls.

McHenry, H. M. and L. A. Temerin. 1979. The evolution of hominid bipedalism: evidence from the fossil record. Yearbook of Physical Anthropology 22: 105-131.

Napier, J. R. 1962. Fossil hand bones from Olduvai Gorge. Nature 196:409-411.

Oakley, K. P. 1968. The Earliest Tool Maker. Sond. aus Evolution unce Hominisation. Stuttlgart: Gustav Fischer Verlag.

Olivier, G. 1976. The stature of Australopithecines. Journal of Human Evolution 5:529-534.

Oxnard, C. and F. P. Lisowski. 1980. Functional articulation of some hominoid foot bones: implications for the Olduvai (hominid 8) foot. American Journal of Physical Anthropology 52:107-117.

Partridge. 1973. Geomorphological dating of cave openings at Makapansgat, Sterkfontein, Swartkrans and Taung. Nature 246:75-79.

Partridge. 1974. Geomorphological dating of cave openings in South Africa, a reply to A.M.J. DeSwardt. Nature 250:683-684.

Pilbeam, D. R. 1969. Tertiary Pongidae of East Africa. Bulletin #31, Peabody Museum of Natural History, Yale University, New Haven.

Pilbeam, D. R. 1972a. The Ascent of Man: an Introduction to Human Evolution. MacMillan Series in Physical Anthropology, New York: MacMillan.

Pilbeam, D. R. 1972b. Evolutionary changes in hominoid dentition through geological time. In: Calibration of hominoid evolution, W. W. Bishop and J. A. Miller, eds., Edinburgh: Scottish Academic Press, pp. 369-380.

Pilbeam, D. R. 1980. Miocene hominoids and hominid origins. Paper presented at the Forty-ninth annual meeting of the American Association of Physical Anthropology, Niagara Falls.

Pilbeam, D. and S. Gould. 1974. Size and scaling in human evolution. Science 186:892-901.

Pilbeam, D., G. E. Meyer, C. Badgley, M. D. Rose, M.H.L. Pickford, A. K. Behrensmeyer and S. M. Ibrahim Shah. 1977. New hominoid primates from the Siwaliks of Pakistan and their bearing on hominoid evolution. Nature 270:689-695.

Prost, J. 1980. Origin of bipedalism. American Journal of Physical Anthropology 52:175-189.

Rak, Y. and F. C. Howell. 1978. Cranium of a juvenile Australopithecus boisei from the Lower Omo Basin Ethiopia. American Journal of Physical Anthropology 48:345-365.

Reed, C. A. and D. Falk. 1977. The stature and weight of Sterkfontein 14, a gracile Australopithecine from Transvaal, as determined from the inominate bone. Fieldiana (Geology) 33:423-440.

Reynolds, V. 1966. Open groups and hominid evolution. Man 1:441-452.

Reynolds, V. 1968. Kinship and the family in monkeys, apes and man. Man 3:209-223.

Robinson, J. T. 1962. Australopithecines and artefacts at Sterkfontein: Part I. Sterkfontein stratigraphy and the significance of the Extension Site. South African Archaeology Bulletin 17:87-107.

Robinson, J. T. 1966. Comment on "The Distinctiveness of Homo habilis." Nature 209:957-960.

Robinson, J. T. 1967. Variation and taxonomy of the early hominids. In: Evolutionary Biology, T. Dobzhansky, M. K. Hecht and W. C. Steere, eds., New York: Appleton-Century-Crofts, pp. 69-100.

Robinson, J. T. 1972. Early Hominid Posture and Locomotion. Chicago: University of Chicago Press.

Robinson, J. T. 1978. Evidence for locomotor difference between gracile and robust early hominids from South Africa. In: Early Hominids of Africa, C. J. Jolly, ed., Duckworth, London, pp. 441-457.

Robinson, J. T. and R. J. Mason. 1962. Australopithecines and artifacts at Sterkfontein. South African Archeology Bulletin 17:87-107.

Sarich, V. M. 1968. The origin of the hominids: an immunological approach. In: Perspectives on Human Evolution, S. L. Washburn and P. C. Jay, eds., New York: Holt, pp. 94-121.

Sarich, V. M. and J. E. Cronin. 1976. Molecular systematics of the primates. In: Molecular Anthropology. M. Goodman and R. E. Tashian, eds., New York: Plenum Press, pp. 141-170.

Sarich, V. M. and J. E. Cronin. 1977. Generation length and rates of hominoid molecular evolution. Nature 269:354.

Schaller, G. B. 1972. The Serengeti Lion - A Study of Predator Prey Relations. Chicago: University of Chicago Press.

Schaller, G. B. and G. R. Lowther. 1969. The relevance of carnivore behaviour to the study of early hominids. Southwestern Journal of Anthropology 25:307-341.

Simons, E. L. 1963. Some fallacies in the study of hominid phylogeny. Science 141:879-889.

Simons, E. L. 1969. Late Miocene hominid from Fort Ternan, Kenya. Nature 221:448-451.

Simons, E. L. 1972. Primate Evolution. New York: MacMillan.

Simmons, E. L. and P. C. Ettel. 1970. Gigantopithecus. Scientific American 227:77-85.

Simons, E. L. and D. R. Pilbeam. 1965. Preliminary revision of the Dryopithecinae (Pongidae, Anthropoidea) Folia Primatologia 3:81-152.

Simpson, G. G. 1963. The meaning of taxonomic statements. In: Classification and Human Evolution, S. L. Washburn, ed., Chicago: Aldine, pp. 1-31.

Sussman, R. L. and N. Creel. 1979. Functional and morphological affinities of the Subadult hand (OH7) from Olduvai Gorge. _American Journal of Physical Anthropology_ 51:311-331.

Steudel, K. 1980. New estimates of early hominid body size. _American Journal of Physical Anthropology_ 52:63-70.

Tattersall, I. 1969. Ecology of North Indian _Ramapithecus_. _Nature_ 221:451-452.

Thompkins, R. J. 1977. Early hominid postcrania and locomotor adaptations. _Kroeber Anthropological Society Papers_ 50:85-104.

Tobias, P. V. 1965. Australopithecus, _Homo habilis_, tool using and tool making. _South African Archaeology Bulletin_ 20:167-192.

Tobias, P. V. 1967a. _Olduvai Gorge, Volume II. The Cranium and Maxillary Dentition of Australopithecus (Zinjanthropus) boisei_. Cambridge: The University Press.

Tobias, P. V. 1967b. Cultural hominisation among the earliest African Pleistocene hominids. _Proceedings of the Prehistorical Society_ 13:367-376.

Tobias, P. V. 1968. The taxonomy and phylogeny of the Australopithecines. In: Taxonomy and Phylogeny of Old World Primates with References to the Origin of Man. _Proceedings of the Round Table at Inst. Anthrop. Centre Primatol_., University of Turin, Italy. Turin: Rosenberg and Sellier, pp. 277-318.

Tobias, P. V. 1969. Commentary on new discoveries and interpretations of early African fossil hominids. _Yearbook of Physical Anthropology_, 1967. S. Genoves, ed., pp. 24-30.

Tobias, P. V. 1971. _The Brain in Hominid Evolution_. New York/London: Columbia University Press.

Tobias, P. V. 1973a. Darwin's prediction and the African emergence of the genus _Nomo_. In: _L'Origine dell Uomo_, Atti del Colloquio Internazionale, Roma Ott., 1971. 182:63-85.

Tobias, P. V. 1973b. Implications of the New Age estimate of the early South African hominids. Nature 246:79-83.

Tobias, P. V. 1976. African hominids: dating and phylogeny. In: Human Origins: Louis Leakey and the East African Evidence. G. Ll. Isaac and E. McCown, eds., W. A. Benjamin, pp. 377-422.

Tobias, P. V. and A. R. Hughes. 1969. The New Witwatesrand University Excavation at Sterkfontein. South African Archaeology Bulletin 24:158-169.

Von Koenigswald, G.H.R. 1973. Australopithecus, Meganthropus and Ramapithecus. Journal of Human Evolution 2:487-491.

Vrba, E. W. 1972. Chronological and ecological implications of the fossil Bovidae of the Sterkfontein Australopithecine Site. Nature 250:19-23.

Vrba, E. S. 1979. A new study of the scapula of Australopithecus africanus from Sterkfontein. American Journal of Physical Anthropology 51:117-129.

Washburn, S. L. 1960. Tools and human evolution. Scientific American 203:3.

Washburn, S. L. 1968. The Study of Human Evolution, Condon Lectures, Eugene, Oregon.

Wells, L. H. 1969. Faunal subdivision of the Quarterrary in Southern Africa. South African Archeology Bulletin 24:93-95.

White, T. D. 1977. New fossil hominids from Laetolil, Tanzania. American Journal of Physical Anthropology 46:197-230.

Wilson, A. C. and V. M. Sarich. 1969. A molecular time scale for human evolution. Proceedings of the National Academy of Science (USA) 63:1088-1093.

Wolpoff, M. H. 1971. Competitive exclusion among Lower Pleistocene hominids: the single species hypothesis. Man 6:601-614.

Wolpoff, M. H. 1976a. Fossil hominid femora. Nature 264:812-813.

Wolpoff, M. H. 1976b. Primate models for australopithecine sexual dimorphism. _American Journal of Physical Anthropology_ 45:497-510.

Wolpoff, M. H. 1979. Anerior dental cutting in the Leetolil hominids and the evolution of the bicuspid P3. _American Journal of Physical Anthropology_ 51:233-284.

Wood, B. A. and C. G. Stack. 1980. Does allometry explain the differences between "gracile and "robust" Australopithecines? _American Journal of Physical Anthropology_ 52:55-62.

Zihlman, A. 1981. Women as shapers of the human adaptation. In: _Woman the Gatherer_, Yale University Press, New Haven and London, pp. 75-118.

Zihlman, A. and L. Brunker. 1979. Hominid bipedalism: then and now. _Yearbook of Physical Anthropology_ 22:132-162.

A SHORT HISTORY OF THE DISCOVERY AND EARLY STUDY OF THE AUSTRALOPITHECINES: THE FIRST FIND TO THE DEATH OF ROBERT BROOM (1924-1951)

Charles A. Reed
University of Illinois
Chicago

Australopithecine is the informal designation for any member of those species which at one time or another, by one author or another, have generally been assigned to the genera _Australopithecus_, _Plesianthropus_, or _Paranthropus_. As often with problems of phylogeny, the ancestry, the descendants, and the duration of different lineages within the group being studied are matters of disagreement, the areas of disagreement changing with time as evidence has accumulated and new investigators enter the particular field of study. These varying disagreements typically have led, with the australopithecines as with many other groups, to a changing pattern of systematics, confusing to students, writers of news articles, and other newcomers to the study.

Our period of knowledge of the australopithecines has been between 1924 and the present (1978, as I write), a total of 54 years. I divide this time into four periods: I. 1924-1951, when the emphasis was on the work and publications of Raymond Dart and Robert Broom, a period of general rejection of the idea that australopithecines were little more than fossil apes; II. 1951-1959, following the death of Broom and until the discovery of the first major fossil australopithecine ('_Zinjanthropus_') at Olduvai Gorge; this period was dominated by the development by Dart of his concept of an osteodontokeratic culture and by a wrangle in England, particularly between W.E. Le Gros Clark and Solly Zuckerman, as to the hominid or non-hominid relationships of the australopithecines; III. 1959-1968, a time of major discoveries at Olduvai Gorge by Mary and Louis Leakey and many arguments as to the phylogenetic position and taxonomy of the recovered fossils; IV. 1968-present, a period of multiple discoveries of rich new australopithecine sites in eastern Africa and of a multiplication therefrom by paleo-anthropologists of contemporaneous lineages of Hominidae.

The whole study began with Raymond Dart. Obviously, if Dart had never existed or had never gone to southern Africa, someone else would have eventually found australopithecine fossils and published on them. But we owe to Dart not only that first recognition of the Taung speciment ('Dart's baby,' as it was soon affectionately known), but also a remarkable intuition, a richness of knowledge and imagination, and a faith in the right of his own analysis which eventually proved triumphant over the doubts and opposition of almost all of his contemporaries.

Dart was born in 1893 in Australia, and received his major education there, culminating with a medical degree from the University of Sydney in 1917. He then served a year in Sydney as a house surgeon in a hospital before moving on to England as a surgeon with the Australian Army Medical Corps in England and France. In 1919, he became senior demonstrator in anatomy at University College, London, a post he held for four years (with an interruption, 1920-21, when he was a Fellow of the Rockefeller Foundation in the United States); during 1921-22 he added the work of lecturer in histology at University College. While these post-doctoral years gave him extensive experience in human and comparative anatomy, he was specializing in comparative vertebrate neurology; his first publication appeared in 1920, and he has continued this activity for more than 50 years.[1] In 1922, at the age of only 29, he reluctantly applied for, and received, an appointment as Professor of Anatomy at the nearly new medical school of the University of Witwatersrand, Johannesburg, in the Transvaal of South Africa. No choice could have served science better. Dart arrived at his new post in January, 1923, and immediately plunged into a program of teaching, administrative organization, and research: during the period of less than two years from his arrival to his first view of a fossil which he was to name _Australopithecus africanus_, he published several papers on neurology and five in anthropology -- three on physical anthropology and two on the distribution and uses of metals in prehistoric South Africa.

The succession of events which led to the discovery of the first and type specimen of _Australopithecus africanus_ has been recounted most charmingly by Dart himself (1967): through the enthusiasm engendered in one of his students, Miss Josephine Salmons, he received two boxes of fossils embedded in limey sandstone of fissure-fills, from quarries in the Gaap Escarpment at Buxton, near Taung, in the Harts River valley of the northern part of Cape Province, South Africa.[2] Opening these boxes, even as he was dressing for a wedding at which he was to be the best man, Dart noted immediately the stony endocast[3] of the right side of the interior of the skull of an advanced anthropoid. From his experience with comparative neurology, Dart immediately recognized that the endocast was much larger than that of a baboon, larger even than that of a chimpanzee, and he believed that additionally and more importantly the proportions of the posterior part of the parietal lobe of the cerebrum were more human than pongid in that the lunate (simian) sulcus was widely separated from the parallel (superior temporal) sulcus. Dart immediately concluded that the owner of that brain had had more parietal associative area than did any known ape.

Thrilled by his discovery, Dart searched avidly for the piece from which the partial endocast had become separated, and was rewarded by finding a large stone with a depression into which the cast fitted perfectly. There was faintly visible in the stone the outline of a broken part of the skull and even the back of the lower jaw and a tooth socket which showed that the face might still be somewhere there in the block (Dart, 1967, pp. 5-6). After that promise of things yet to be revealed the wedding was a time-wasting hindrance to which he was finally dragged away by the groom.

Dart has stated (1967, p. 10) that his discovery, and the date of his friend's wedding, was 73 days before December 23, 1924, thus on or very close to October 11. Robert Broom (1946, p. 12) stated that Dart received this specimen in November 1924, but Dart's own memory is probably more accurate.

Lacking any paleontological experience, yet convinced of the great importance of his find, Dart set himself the task of removing the rock from around the face, using with great delicacy a small hammer, a small chisel, and finally one of his wife's knitting needles. He worked every second he could spare from his numerous duties, until, two days before Christmas of that same year, he had the face (including the lower jaw) exposed. What he saw was the facial skeleton of a young child with the deciduous dentition present and in an advanced state of eruption, so that the age was equivalent to that of a modern human of six years. Also present, in addition to the human-like characters of the endocast, was a relatively high forehead and a muzzle less protruding than for any ape of equivalent age. Lacking was any evidence of brow ridges, already present in young apes of the same age, and most noteworthy was the absence of projecting and interlocking canines, obvious characters of the deciduous canines of young chimpanzees and gorillas.

Dart had a unique fossil, and he knew it; he had drawings and photographs prepared, and wrote a preliminary article of some 3500 words, which he sent by steamship to NATURE, in London, where it arrived in time to have proofs set up and distributed to several British scientists by February 3, 1925. On that same day, an announcement of the find was published in the STAR, a newspaper in Johannesburg, and cables of that report went out to all the world. The next day was Dart's 32nd birthday, and on February 7 his article was published in NATURE.[4]

In this article, Dart presented his major conclusions, mentioned above, that the partial skull and endocast from Taung represented a young individual of a population which possessed many hominid characters and indeed might have represented a stage in human evolution. Even so, he continued cautiously to call his 'baby' an ape, but one so distinct from all other known apes, fossil or living, that it probably deserved to be placed in a new family, the Homo-Simiidae. He also proceeded to deduce that, because of the projected position of the foramen magnum (a deduction based upon the configuration of the endocast and a small adhering remnant of an exoccipital) the individual probably stood and walked erect, thus freeing the hands for non-locomotor activities. He then bestowed a new generic and a new specific name, *Australopithecus africanus*, the 'southern ape of Africa,' upon his discovery.

The reaction was immediate, four answering letters in the next issue (Feb. 14) of *NATURE*, one each by Sir Arthur Keith, and Professors G. Elliot Smith, Arthur Smith Woodward, and W.L.H. Duckworth plus an additional article the same day by Keith (1925b) in a different journal. Woodward was the most adverse, Duckworth the most favorable, but none accepted Dart's evidence for the hominid nature of his new 'ape' and none (rightfully so) accepted his new family, but the general opinion was that the fossil represented only a young (and therefore non-diagnostic) individual of a population probably related to chimpanzees or gorillas.

The reasons, both published and unpublished, were in part profound and in part trivial, as follows: 1) the obvious speed with which Dart had reached his conclusions and prepared his manuscript presumably could not have allowed him time for mature judgment; 2) he lacked comparative specimens and an adequate library; 3) an early hominid-like fossil should, like Piltdown Man, have had a large brain but primitive jaws and dentition, with large canines; 4) Dart's claims for hominid-like functions of the brain were insecure; 5) the evidence for bipedal locomotion was weak[5]; 6) Africa was thought to be the wrong continent for an early hominid fossil, since at that time Pithecanthropus from Asia was the only known 'ancient' fossil hominid other than the Neanderthals, Piltdown Man, and the Mauer jaw (Heidelberg Man) from Europe; 7) A juvenile skull gave insufficient basis for a comparative analysis, since young apes and young humans are well known to be more similar to each other than are adult apes and adult humans; 8) Dart's literary (sometimes lyrical) and thus non-scientific style of writing and his zealous enthusiasm for his cause produced a negative reaction in some; 9) The creation of a new family was unwarranted on the basis of a single, juvenile specimen; 10) Did Dart perhaps think he was a second Dubois, that he could find a new hominid ancestor in an obscure part of the world so soon after his arrival there?; 11) Dart, as a post-doctoral student in England, was known to be brilliant but impetuous; 12) The etymology of the generic name was improper, a bastardized mixture of Latin (australo) with Greek (pithecus); 13) Dart did not mention anything about the geology of the site or the geological age of the specimen so its place in time was unknown, thus a supposed ancestral position could not validly be deduced; 14) None of the critics had seen the specimen nor were casts available,[6] and indeed each may have had the thought that the specimen should have been submitted to scientists in Britain for detailed study by experienced specialists before anything was published by the young discoverer; 15) The extremely small size of the illustrations published in Nature truly did not allow independent verification of Dart's anatomical descriptions.

To add to the weight of the adverse criticism, A. Robinson, a professor in Edinburgh, compared Dart's pictures with skulls of four chimpanzees available to him, and declared that some of those four differed more from others of the same group than did Dart's specimen. In other words, in his opinion Dart had described a young chimpanzee.

Considering the weight of the attack, Dart was remarkably restrained, responding at first (1925b) only to the criticism of his etymology, pointing out that numerous Greek words (including pithecus) had been used as Latin by Roman authors. In July, however, Keith (1925b) returned to a fray of his own making with a much more intensive criticism of Dart's ideas than he had expressed in February. Dart responded to this assault, he and Keith publishing adjoining letters in NATURE for 26 Sept., 1925; each author offered new arguments which failed to have any influence upon the other. Indeed, Keith was more clearly opposed in this, his fourth publication of the year on Australopithecus, to Dart's find having any particular hominid affinities than he had been in February.

Dart gained only three supporters that year, Professor W.J. Sollas, Professor G. Elliot Smith, and Dr. Robert Broom. The first two of the group were eminent British scientists living in England, but neither -- in spite of their early and continuing support of Dart -- seemingly had much influence on others' opinions, and thus are not of great importance in the history of the australopithecines. The situation with Robert Broom is vastly different; for several years, as we shall see, he was much more active than was Dart or anyone else in finding new australopithecines and writing about them.

Robert Broom, born in 1866, was a Scotsman from the area of Glasgow; the family was poor but his father read widely. Broom's early schooling was meager, but he had a tremendous interest in all aspects of natural history, learning as he could from books and from other amateurs. In spite of having had only four years of formal schooling, at 16 he entered the University of Glasgow, majoring in chemistry and working as an unpaid laboratory assistant and demonstrator. Although the generally poor teaching he received did not stimulate him to become a chemist, his first scientific paper (and his only one in chemistry) was published while he was an undergraduate. At the age of 20 he received his B.Sc., with honors in Geology, and in 1887 entered the Medical School of the University of Glasgow, from which he emerged two years later as a Bachelor of Medicine and a Master of Surgery. He could practice medicine but he was not a 'doctor.'

Too restless to settle down but too poor to pay for traveling, he twice used his medical degree to get positions as a ship's physician on small steamers taking immigrants to the United States, so saw a little of North America's eastern coast and visited the museums at Harvard, and others in Boston and Philadelphia. Next, again as a ship's physician, he went to Australia, where he stayed for four years. In Australia he collected avidly, discovered a site of fossil marsupials, did much comparative anatomical and embryological research with the simplest of equipment, and supported himself in several small settlements by practicing medicine at phenomenally low pay, a pattern he was to follow later in South Africa. Working in various parts of Australia's outback with no scientific library, he sent requests for articles to his father, who, using libraries in Edinburgh, copied in entirety by hand (including drawings and some articles in German) the articles requested. By the time Broom left Australia in 1896 he had published 26 articles, and more appeared later from the work done there. One such article finished in Australia was on the comparative anatomy of the Organ of Jacobson; submitted as a thesis to the University of Glasgow, he was awarded the M.D., and henceforth could call himself 'doctor.' Also while he was in Australia, his sweetheart, Mary Baird made the long trip from Scotland, and they were married; her life thereafter was never an easy one as Broom sacrificed everything to his research, but she was never heard to complain and was with him when he died 58 years later.

Broom's anatomical research on monotremes and marsupials stimulated an interest in the evolutionary origins of mammals which was never to desert him, but at the same time he had become convinced that one could not solve these problems only by research in comparative anatomy and embryology; one also had to study the fossils of the animals which had undergone the transition. Some of those fossils had been found in the Karroo (Permian-Triassic) strata of South Africa, so, after a brief visit home, the Brooms went to South Africa, where they lived the remainder of their lives. For the most part, Broom collected his own fossils, was his own laboratory technician, sketched his own illustrations with the same pen he used for writing, had no library but his own, and most of the time supported himself, his family, and his research by the practice of medicine in small towns which happened to be near where he was finding fossils of the mammal-like reptiles he was studying. Under these Spartan conditions, by incredibly hard work and simple brilliance, he became the leading student of the problem of mammalian origins, and one of the most famous paleontologists in the world. He also became a philatilist, continued with his studies of comparative anatomy and embryology, and had other divertissements.

Broom was immediately fascinated by what he read of Dart's new fossil and decided to see it. He wrote ahead of his coming and then went to Dart's laboratory in Johannesburg for a weekend of intensive study of the new specimen, _Australopithecus_, during the latter part of the same month (February 1925) of Dart's original announcement. Broom at the time was 58, world famous, and already author of almost 250 publications. He was also a master of intensive concentration as based upon a tremendous knowledge of the structure and evolution of vertebrates. Broom studied Dart's fossil, whole-heartedly agreed that it represented a population which had been more hominid-like than any ape hitherto known, and promptly prepared an article for NATURE (published 18 Apr., 1925), stating his support for Dart. This short article, illustrated with five of his own quick freehand sketches, as were all of his articles, mostly fell on deaf ears. Broom was familiar, as Dart was not at that time, with the geology of the area from which the fossil had come, and presented the first short description of that topic for a general audience, but in so doing did a considerable disservice to acceptance by British paleo-anthropologists of any idea of _Australopithecus_ being related to hominids by stating that, although the geological age of the fossil was not determinable, it was probably not older than the Pleistocene and could be as recent as Rhodesian Man. This state of possible chronological position, coming from a person with Broom's high reputation as a geologist,[7] was generally accepted as evidence that _Australopithecus_ must have been contemporaneous in time with unknown hominids already well-evolved toward the living type; indeed Keith, in his arguments later that year, used Broom's estimates of time exactly in that way.

A second article (Broom 1925b) prepared at the same time contained more detail on the geology of the area and on comparative osteology. Broom sent the manuscript to the American Museum of Natural History, where as a paleontologist he had worked in 1913, and it was promptly published (April 1925) in NATURAL HISTORY. In this article, Broom not only presented an accurate drawing of what an adult of the Taung specimen should look like, but also included drawings of the cusp patterns of the upper and lower first molars, even though the occlusal surfaces of these teeth were not to be exposed until four years later. His guess that the time at which the fossil lived could have been between 10,000 and 100,000 years ago illustrates not only the ideas of geologists of the period about time, but also that even the best of scientists, lacking all hard data, should be cautious in their pronouncements.

This second article by Broom seems to have been almost completely overlooked in most of the subsequent history of the study of australopithecines, although he himself referred to it in his monograph of 1946 (p. 17). Even so, H.B.S. cooke, a paleontologist who had done much research on the mammalian fossils from South African australopithecine sites, could write in 1960 that, until that date, he had not seen (and obviously had not known) Broom's article of 1925 in NATURAL HISTORY. Additionally, this article is rarely mentioned by anyone writing on australopithecines and obviously was rarely read.

The American naturalist Frederic A. Lucas (1925) quickly put an end to the etymological problem concerning _Australopithecus_ by reminding everyone of what they had forgotten in their zeal for the niceties of the usage of Latin and Greek; the International Rules of Zoological Nomenclature require only that a proposed new generic name consist of a unique combination of letters, so etymology is not involved. At the same time, he pointed out that a new name for a family should be based on a generic name, so Dart's 'Homo-Simiidae' was invalid.

As mentioned above, two English authorities, in that summer of 1925, did change their views and swing around to agree with Dart and Broom. Professor W.J. Sollas, geologist and anthropologist at Oxford, had long been a correspondent of Broom and asked him to send a drawing of the mid-sagittal section of 'Dart's Baby,' which Broom did. Comparing this section with those prepared from skulls of young apes in the collection at Oxford, Sollas soon realized that _Australopithecus_ was much more human than pongid, and he published these conclusions, illustrated with the comparative sections in two articles (Sollas 1925, 1926). He did, however, state that an adult skull was a necessity for a final decision as to the place of *Australopithecus* in human phylogeny. The first paper by Sollas, plus the weight of Dart's and Broom's observations, had the effect of converting the neurologist and physical anthropologist Grafton Elliot Smith, who had been Dart's professor at University College, London, to the view of the hominid-like nature of *Australopithecus*. Another who spoke favorably, in the comments following Sollas' second publication (p. 10), was the paleontologist D.M.S. Watson, who observed that the skull "...was different than that of all other great Apes, yet wonderfully human in appearance, in dentition, etc. The brain-cast had impressed all neurologists by its very human appearance. The Taungs skull was that of an Ape more like Man than any other Ape yet known."

So then there were five who believed-- Dart, Broom, Sollas, Smith, and Watson. (Perhaps Watson was only favorably impressed.) This was an eminent group of scholars, but Smith and Watson were never active in support, nor was Sollas after he published his two articles, and so in England interest languished and the opinion persisted that Dart's ape, although different from other apes, was no more than another ape.

In South Africa, meanwhile, the American physical anthropologist Ales Hrdlička visited Dart in August 1925 to look at *Australopithecus*. He was impressed with its unique character, thought it represented a new species and perhaps a new genus of ape, but he was not so favorably impressed with its supposed hominid-like characters. That same year (1925), Hrdlička published his observations on the fossil and on the geology of the site as seen by himself on his visit to the Buxton quarry from which it had come. His one new anatomical contribution was that the enamel was 'high' (thick), as in humans, but no one followed this lead for several decades. Hrdlička's account also testified to his own agility and persistence; he was 56 at the time but spent several hours, slung by a rope on a vertical face of a cave-filling, trying vainly to chisel out some bones of fossil baboons. His published expectation that when the baboons were identified they would give a clue to the age of the deposits was unfortunately not fulfilled, but his concluding comment that "The antiquity of the find is not necessarily very great..." could be interpreted easily to indicate an agreement with Broom of the relative recency of *Australopithecus*, when actually at the time no evidence as to age was determinable beyond the fact that the fossils were more recent than the time of formation of the tunnels and caves which had formed in the limestone being quarried and the same age as the sandy-lime deposits which subsequently filled those caves and tunnels in the limestone. Since no absolute age was known for the limestone, for the time the caves had been formed in them, or for the time the caves had later been filled, no logical guess could be made at that time for the age of any contained fossil. Actually the guesses that were forthcoming within a few years ranged from Early Pliocene to no older than a few thousand years. Yet Keith (1925d) and others discounted the possible ancestral hominid status of *Australopithecus* on the basis that it was so recent that it was possibly no older than the newly-discovered Rhodesian Man (whose age was also unknown), whereas the hominid lineage (the 'phylum of man') had supposedly separated from that of the apes in the Miocene. Obviously, if *Australopithecus* was only a few tens of thousands of years old, it could not be a hominid ancestor.

After leaving South Africa, Hrdlicka gave a lecture in England on *Australopithecus*, in which he expressed his honest uncertainties as to its place in anthropoid phylogeny and taxonomy, and stated (as others had) that a skull of an adult would be needed to settle these problems. In his monograph of 1930 on the skeletal remains of early man, *Australopithecus* was not mentioned, and in writing on *Ramapithecus* in 1935, Hrdlicka stated that this latter, while an ape, yet was more manlike than any other ape, including *Australopithecus*. Krogman (1976) has stated that Hrdlicka was an early supporter of Dart and Broom in their belief that *Australopithecus* was hominid, but such support, if true, must necessarily have been after 1935.

During that same year, 1925, of the announcement of *Australopithecus*, new boxes of fossils were being sent to Dart, including some from limeworks at places named Sterkfontein and Makapansgat (Dart 1925d)[8] but the samples lacked evidence of baboons or anything resembling *Australopithecus*; instead they contained mostly pieces of antelopes. Dart at that time had little time for fossil antelope, and so his interest waned. Additionally, in that year of 1925, Dart was made Dean of the medical school at the University of Witwatersrand and was elected president of the anthropological division of the South African Association for the Advancement of Science. These were singular honors for a young man of 32, who at the beginning of the year had been almost totally unknown, but at the same time his responsibilities as dean, while retaining his position as head of the Department of Anatomy, increased his administrative burden tremendously. The medical school, which at the time did not even have a library, had already budded off a dental school, and for four years Dart was also dean of that school.

Dart was criticized for the brevity of his two articles of 1925, so the following year (Dart 1926) he prepared a longer and more popular account in which he did not dwell upon anatomical detail, which as necessary he showed in pictures and in three diagrams of lateral views of skulls of a young orang, a young chimp, and a young gorilla successively superimposed upon the same view of the type of *Australopithecus*. Both chimpanzee and gorilla were quite different from *Australopithecus africanus* but the chimp was less different. The major emphasis of the article was on the environment of southern Africa (which at the time was considered to have remained relatively unchanged since the Cretaceous), the impossibility of any typical ape surviving in such an environment, and the necessity of erect, bipedal locomotion for any primates of the size of adult *Australopithecus* to survive in such an environment. He wrote further about the relative sizes of brains of living apes and of his juvenile specimen, and about the presumed mental powers of *Australopithecus* relative to the size and configuration of the endocast and the presumed selection-pressure on a bipedal ground-living primate in the environment of South Africa. He believed his 'baby' had belonged to a cave-living population (a view he has never abandoned), and that that population had preyed upon many of the smaller animals, including possibly baboons, whose bones were also found in the limey cave-fillings of the Gaap Escarpment. He wrote about the possibly Pliocene or pre-Pliocene time during which *Australopithecus* had lived, but no evidence was available at the time by which to specify a date, and, as stressed before, Broom (1925a) had already set the intellectual mold with his statement that the skull was ". . .not older than Pleistocene and perhaps even as recent as the *Homo rhodesiensis* skull." This statement of course negated everything Dart had written and supported the critics who claimed that *Australopithecus* was too recent to represent a link between any ape and Pithecanthropus, the earliest hominid fossil then known. (Actually, little was known about the dates of any hominid fossils at the time, either relative or absolute, nor do we know yet the absolute dates of the australopithecine fossils from South Africa.) Lastly, Dart repeated that his 'baby' represented a new family, intermediate between pongids and hominids, and no one else believed that.9

In this publication of 1926, longer than any on Australopithecus previously by Dart, he failed to bring in the supportive evidence already published by Broom and by Sollas; instead he seems to have assumed in part that his readers of an American journal would be familiar with the facts, and he discussed his own deductions. Several of these have proved to be correct, but some were not; in any case this interesting article of 1926 by Dart seems to have had little impact and was rarely mentioned in later literature on Australopithecus.

During the four years after 1925, in addition to being a teacher and administrator and busying himself with other research, Dart found a few minutes now and again to continue the preparation of the skull of his 'baby,' a difficult and tedious labor of removing the limy deposit between the upper and lower teeth so that the jaws could be separated and the occlusal surfaces studied.

Broom, ever impatient, could not wait for the final separation of mandible and maxillae. He had already (1925b) conjectured and drawn the occlusal surfaces of the first molars, so in a later study of 1928 he concentrated on the deciduous premolars. From what he could see in lateral view, he prepared a description of the cusps, complete with drawings of the occlusal surfaces! This manuscript was received in London on 11 Dec., 1928, and the final separation of the mandible did not occur until 10 July 1929. Broom's sketches, by then already published (Broom 1929), were quite accurate; the cusp-patterns of the deciduous premolars actually were more similar to those of humans than to those of a chimpanzee or gorilla, quite as he had stated.

During these same years, 1925-1929, Dart had been writing a definitive monograph on both the endocast and the skeletal part of his 'baby,' and now added a description of the teeth. From the time of the first announcement of *Australopithecus*, paleontologists and anthropologists had been waiting, sometimes impatiently, for a complete description and analysis of the fossil, but now that monograph was finished. Dart expected speedy publication, but circumstances decreed otherwise; except for the part on the dentition, the monograph remains unpublished to this day!

The British Association for the Advancement of Science met in Johannesburg in July 1929, and Dart expected great interest to be shown in *Australopithecus*, which he made the theme of a special display. However, in Dart's chapter on Anthropology in the Handbook prepared for the meeting (Crocker and McCrae 1929) he did not emphasize *Australopithecus*, indeed in my opinion he under-emphasized it in comparison with his treatment of many other aspects of anthropology in South Africa and the Rhodesias. Dart was disappointed in the generally apathetic reaction to his display; seemingly almost everyone considered the specimen from Taung to be merely a fossil ape and thus not worthy of special attention. The American paleontologist A.S. Romer did examine the fossil, but then published only a short note (1930) stating that, whatever the young *Australopithecus* was, it was not a chimpanzee, as some had thought.

Dart then resolved to submit his monograph for publication to the Royal Society in London, and to visit there himself, with his prized specimen in hand. He had had photographs and casts of the upper and lower teeth prepared, and mailed these to dental experts and physical anthropologists abroad. He also mailed his monograph ahead, to Professor Elliot Smith in London, sent his precious 'baby' by sea with his wife, and himself took eight months to travel overland the length of Africa on a well-earned vacation.

In the meantime, Broom had been having second thoughts about the age of *Australopithecus*. By that time the fossil baboons from Taung had been studied and classified; since they were all extinct species, Broom postulated they must have lived much longer ago than he had guessed at first. However, his small note in NATURE (Broom 1930) stating these facts and then estimating the age of the fossil baboons and of *Australopithecus* as Pliocene, and possibly early Pliocene, was not sufficient to reverse the effects of his earlier statements as to the *late* dating for

Australopithecus, particularly as his new estimate, again, was no more than another guess.

The next publication on Australopithecus, one which should have carried much weight because of the high reputation of its author, William King Gregory, appeared in SCIENCE of June that same year (1930). Gregory was a vertebrate comparative anatomist and paleontologist of wide experience whose outstanding research on the evolution of Primates was but a small part of his total productivity. Within the field of physical anthropology he had studied many fossils, was the main defender of a dryopithecine, and thus arboreal, ancestry of hominids, and was also the primary authority on the evolution of the cusp-patterns of the molars. His book of 1922, "The Origin and Evolution of Human Dentition," is still the necessary foundation for any dental anthropologist today.[10]

Gregory had received photographs of the occlusal surfaces of Australopithecus and, on the basis of 26 characters he used for comparisons, had constructed the following, oft-quoted table:

Nearer to chimpanzee	0
Nearer to gorilla	2
Nearer to chimpanzee and gorilla	1
Common to chimpanzee, gorilla, Australopithecus, and primitive man	3
Transitional to, or nearer to primitive man	20
Total	26

Gregory completed his article with the following question and statement: "Now in the light of all this additional evidence, if Australopithecus is not literally a missing link between an older dryopithecoid group and primitive man, what conceivable combination of ape and human characters would ever be admitted as such?....Australopithecus, to judge from its skull and dental characters, was a pioneer in the new line (i.e., bipedalism), as held from the first by Dart."[11]

Dart was a week into his African trip when Broom's article on an older age for Australopithecus was published in England and he was a month away from his desk when Gregory's statements were published in the United States, but somewhere along his journey he must have heard of these events, and been heartened, but neither article seems to have made much, if any, impression on Europeans. For one thing, Gregory's statements about *Australopithecus* was tucked in, as if an afterthought, at the end of a longer article on general problems of human evolution and so may have been missed by many, but the major factor seems to have been that, by 1930, the scientific world was excited about some obviously recognizable ancestral humans, the fossils of *Sinanthropus pekinensis* from Choukoutien, near Peking, and people were no longer interested in 'apes' from South Africa, nor cared whether they were early Pliocene or late Pleistocene, nor cared even what as eminent a scientist as Gregory thought about them. Of these currents of thoughts in Europe Dart know little as he wandered his way the length of Africa.

Dart arrived in London early in February 1931, expecting the best, but met with bitter disappointment. Unknown to him, Wolfgang Abel, then an assistant in the Anthropological Institute of the University of Vienna, had published that same month a detailed article of 100 pages of descriptions of *Australopithecus africanus*, based on commercial casts[12] only recently prepared in London and sold to the Paleontological and Paleobiological Institute of the University of Vienna. Abel had gathered from the museums of central Europe a considerable number of skulls of great apes and of human children for comparative purposes, and he also had the extant literature on *Australopithecus* for his use; he is one of the few I have seen to quote Dart (1926), but he failed to quote Broom except for his first article (1925a) and he did not mention Sollas. Lastly, he had had the opportunity of viewing the original type specimen when Mrs. Dart, while waiting for her husband to emerge from Africa, stopped in Vienna and gave an exhibition of the specimen to an informal gathering of Austrian scientists (Dart 1967, p. 57).

While the lack of ethics of publishing on another man's research materials without a word to him on the matter should not pass unmentioned, Abel had done a thorough study of endocast, face, and dentition, with measurements, photographs, and drawings. His conclusion agreed with those already made in 1928 by his father, the renouned vertebrate paleontologist Othenio Abel, "O. Abel sieht in *Australopithecus* einen jungen Gorilla"[13] (W. Abel, 1931, s. 551). Not everyone in England had read Abel's hundred pages of German (nor have I), but everyone knew his conclusion (totally erroneous as time has proved): Dart's 'baby' was a young ape related to an ancestral gorilla![14]

Additionally, Keith had a book in press (published that same year, 1931), in which he described and discussed Dart's specimen at some length (42 pages), and concluded again, as he had done so often in 1925, that Australopithecus was a fossil ape related to chimpanzees and gorillas, but more similar to the chimpanzee. Keith recognized some of the hominid-like characters Dart had stressed, although he was not convinced of the proper identification of the parallel and lunate sulci crucial to Dart's interpretation of the functions and evolutionary status of the brain. Keith interpreted the several hominid characteristics as due to independent evolution in a population of apes which had had a common ancestry with chimpanzees and gorillas.

The differences between Dart and Keith were mostly matters of emphasis. Dart had never regarded his 'baby' as other than an ape, although claiming it as an ape with such remarkably human-like characters as to be between the lineages of known pongids and known hominids; Keith considered Australopithecus to have been an extinct side-line of apes, showing some few human-like evolutionary tendancies but still within the general group of chimpanzees and gorillas. From Dart's point of view, Keith interpreted the evidence negatively, whereas from Keith's point of view Dart had over-emphasized what seemed to be a relatively few hominid-like characters and had tended to weight too lightly all the more pongid characters. With regard to Dart's claim for bipedalism, Keith found no paleo-environmental evidence to support Dart's claim that southern Africa at the time that Australopithecus lived there had not been forested; indeed, Keith reasoned, if a population of apes was living there, forest must have occurred. Keith also pointed out that a young chimpanzee with its first molars newly erupted has its foramen magnum in a forward position under the skull, quite as Dart had described that of his young Australopithecus, but as the animal matures the skullbase grows differentially and the position of the foramen magnum becomes more posterior; he saw no reason not to assume that the same would happen during growth to maturity of an Australopithecus. Keith did not quote Sollas or refer to the evidence published by the latter in support of Dart's opinions. Keith's major error was in contrasting the specimen of *Australopithecus* with children of modern races as if there had existed no possible intermediary hominids; naturally he found major differences, which he then used to discount the possibility of *Australopithecus* being an early hominid ancestor. This error, however, was one not detected by his contemporaries.

One other factor worked against Dart in 1931; in England, as I have already mentioned, the discoveries at Choukoutien of *Sinanthropus pekinensis* (now termed *Homo erectus pekinensis*) had all the British paleo-anthropologists excited, and Dart's presentation of his specimen and his conclusions to a meeting of the Zoological Society of London (17 Feb. 1931) was a dismal anticlimax, in his own opinion, to a far more interesting discussion of *Sinanthropus* earlier the same evening by Elliot Smith. The Royal Society, considering that Abel and Keith were already publishing so much on *Australopithecus*, offered to accept only Dart's description of the dentition, but declined his monograph in its entirety. Dart refused this incomplete approach, and returned to South Africa with his monograph. The fact that he and his few supporters were right and all others wrong was little consolation at the time; Dart's monograph remains unpublished, and indeed an adequate description of the type specimen of *Australopithecus* has not yet appeared.

Refused publication by the Royal Society and rebuffed intellectually, even if accepted socially, by his peers in England, Dart returned to South Africa a little discouraged but his questing spirit and boundless energy soon led him into numerous researches not connected with australopithecines, and papers on multiple topics continued to tumble out, all through the years of the Great Depression. One bright ray of light, appearing soon after Dart's return to Johannesburg, was the publication in 1932 by Professor P. Adloff of Konigsberg of a study he had made on the dentitions of *Australopithecus*, apes, and humans. Adloff's own conclusion agreed with that of Gregory and was totally opposed to Abel's pronouncement of the previous year. "Das Begiss von *Australopithecus* ist aber *rein menschlich und lasst den Schluss zu, dass Australopithecus kein Anthropoide, sondern ein Hominide ist.*" (Italics his). Adloff wrote that the dentition was human, and otherwise indicated that the animal had been too; he was the first to be so definite in using the word 'hominid,' but his view was extreme for the time, and not even Dart or Broom or anyone else for several years showed such courage, but instead continued to regard *Australopithecus* as an 'advanced' type of ape.

In 1934 Dart published a short and somewhat general paper on the dentition of *Australopithecus*; he obviously regarded the paper as a preliminary one. Although sumitted in 1933, the manuscript could have been (and indeed may have been) composed some years earlier. It lacks a bibliography, and neither Abel (1931) or Adloff (1932) were mentioned; perhaps Dart considered their studies, necessarily made from casts of the dentition of *Australopithecus*, as not definitive. Dart not only had the advantage of the original specimen, but he had x-rays of the jaws, showing the unerupted permanent dentition. Other than five brief sentences, he did not however utilize the evidence seen in the radiographs, reserving that material for the more complete monograph he was still planning. He did mention one important fact: "The permanent canines like the milk canines are humanoid." Dart's final conclusions were essentially in agreement with those of Adloff: the dentition of *Australopithecus* was distinctly human, and if any one of the teeth had been found separately it unquestionably would have been assigned to a man. Dart did not, however, go as far as Adloff, but spoke instead of "...the dentition of a man in the jaws of an ape." On the basis of the canines and incisors in particular, he favored *Propliopithecus* of the Oligocene of the Egyptian Fayum as the ancestor of the "Australopithecidae," thus bypassing the dryopithecines, favored by Gregory as hominid ancestors, for both *Australopithecus* and hominids.

I cannot see that this article of Dart's, perceptive as it was in many ways (while at the same time also containing what we now regard as errors) had any immediate influence toward changing anyone's mind about the hominid-like nature of *Australopithecus*. Seemingly in the mid-1930s no further information which could be wrung out of the single specimen was going to have any influence on anyone at that time. More specimens were required and these were soon to appear.

The emphasis of our history now shifts to Dr. Robert Broom, who in 1928 attempted to limit his medical practice to give himself more time for research. He did well with the research, producing three books and numerous articles, but fell behind financially. Returning to medical practice in 1932, probably the worst year of the Depression, was no help, and the following year he was so poor that, although president of the South African Association for the Advancement of Science, he could not have attended the annual meeting if the Association had not paid his expenses. Dart had become aware of Broom's problem, and in 1933 he wrote to Field-Marshall J.C. Smuts, then Prime Minister of South Africa, suggesting a government position for Broom. Dart never received an answer, but the following year Broom, then 67, was offered a temporary position as Assistant for Paleontology and Physical Anthropology at the Transvaal Museum, Pretoria. The title was an insult, for one of history's greatest paleontologists,[15] but the salary -- slightly more than 40 Pounds a month when the Pound was worth $4.80 and the dollar was worth ten times what it is today -- was more than adequate[16] during a world-wide depression, and was almost three times what he had been making in private practice. Additionally, the release from medical practice gave him time for sustained work on reptilian fossils, but after preparing 16 papers in 18 months, his thoughts began to turn to the possibility of finding additional specimens of *Australopithecus*, or possibly other primates, or even other Pliocene or Pleistocene mammals. Broom was always eclectic; at the same time he might be thinking and/or writing on flight in the ancestors of ostriches, on Permian amphibians, on the milk-dentitions of living large felids, on rare postage stamps, on fossil hominids and their phylogeny, and several other topics. (He also had a keen eye for pretty ladies, and for good paintings offered for sale at low prices.)

The area west of Pretoria had numerous lime-filled caves and fissures, and several of these (as at Taung) had fossils. Broom sampled some of these with considerably success, but at first found nothing like *Australopithecus* until he was taken to a commercial limeworks at Sterkfontein, 40 miles southwest of Pretoria and thus nearer to Dart in Johannesburg than to Broom. Rich concentrations of fossils had first been described at this site in 1897 (Broom, 1950, p. 41); in the intervening years a few had been saved, but most had been shoveled into the lime kilns. Broom was led to this site by two of

Dart's students, Harding la Riche and G.W.H. Scheppers, who in turn had been shown the caverns by the zeal of another student, Trevor Jones, who was writing a thesis on skulls of extinct baboons he had collected there. At Sterkfontein, Broom met the man in charge of quarrying, G.W. Barlow, who had been collecting old fossils from the cavern's fillings and selling them to tourists. Barlow had worked at Taung, remembered Dart's famous skull there, and said he would keep a lookout for anything like it.

Eight days later, August 17, 1936, Barlow handed Broom the anterior two-thirds of an endocast similar to that of the specimen from Taung, the first such specimen of *Australopithecus* to be recovered for science since 1924. This endocast had been discovered after blasting, the standard method of commercial quarrying but not a good way to salvage intact fossils. No additional parts were found that day, but the next day careful searching by Broom and several helpers turned up the base of the skull, portions of the parietals and frontals, and blocks of breccia from which the specimens had been shattered. From this breccia, after several weeks of work at the Museum, were retrieved the crushed maxillae, most of the bones around the orbit, and several teeth. Thus, in nine days after his first visit to Sterkfontein, Broom had recovered a considerably part of an adult australopithecine skull, in contrast to all of the prior years of commercial mining, during which the fossils had been burned or sold as curios.

With his usual swift efficiency, Broom had two manuscripts in the mail for London on August 28th; both were published on September 19th, a popular account in the ILLUSTRATED LONDON NEWS and a more technical article in NATURE. In the latter, Broom (1936a) named his new fossil *Australopithecus transvaalensis*,[17] choosing not to use Dart's name *africanis* as a species on the logical basis that the differences between the skull of a child (Dart's baby) and that of the new adult did not allow him to know that both belonged to the same population. Later, on the basis of the same logic but not so well-founded at the generic level, Broom (1938) changed the generic name to *Plesianthropus*, a name generally forgotten today, but Broom was of a generation when differences were noted more than similarities and, particularly in paleo-anthropology, each new specimen would receive its own generic and specific names.[18]

The population from Sterkfontein -- and I write 'population' purposely, because new specimens continued to be found, cleaned of matrix, catalogued, drawn, described, and published -- were what we now call 'gracile australopithecines.' Neither Broom nor Dart had any doubt of their affinity to the single specimen of the child from Taung, nor did any other paleontologist. Thus *Australopithecus africanus* came naturally to be considered a gracile australopithecine. By June of 1938, when Broom first found and differentiated a second and more robust type of australopithecine as *Paranthropus robustus*, the affinity of 'Dart's baby' had been automatically and firmly settled with the gracile type. If Broom had found robust individuals first and his graciles later, I believe the opposite situation would have developed as automatically, and the child from Taung would today be considered a juvenile of the robust group, and the latter named accordingly. Since the affinities of Dart's type specimen of *Australopithecus africanus* with the robust population may be the true one (Tobias 1973a,b) it is perhaps unfortunate that Broom did not find Kromdraai first, and Sterkfontein later!

The taxonomic problems were far in the future, however, when Broom and his helpers from the Transvaal Museum were first collecting at Sterkfontein. Dart was interested in the progress of the work, but he had never regarded *Australopithecus* as his primary interest; additionally he was involved with administrative detail and with other research. The two men easily kept in contact, however, as from 1934 into 1940 Dart engaged Broom as a part-time lecturer, two mornings a week, to speak on random topics to the advanced classes at the medical school. Broom easily made the round trip twice a week by train, but one wonders how much influence he had on the medical students, most of whom (as true generally for medical students) were hard pressed with their immediate studies and clinical activities and were little interested in attending lectures on geology, evolution, fossils, embryology of Australian mammals, botany, or whatever else came to Broom's active mind, but a few were charmed. One such was a student who would sometimes ride with Broom on the train, to listen more, and who later wrote Broom's biography (Findley 1972).

During 1936-37 the hominid similarities, particularly of the teeth (Broom 1936; 1937a, b, c; 1938b) of the australopithecine fossils collected at Sterkfontein were becoming more apparent as new specimens -- bits of broken jaws, pieces of broken skulls, a few fragments of long bones, or sometimes only isolated teeth--appeared and were studied. However, all the specimens were still regarded as having been derived from apes, or were sometimes called 'ape-men,' the latter a usage which, although now must be considered quite erroneous, is still met (sometimes by people who should know better).

Broom had his 70th birthday on November 30, 1936. This is an age when most people have retired or soon will, but Broom's typical active life was to stretch on into the future for almost 15 years. Of his total scientific publication of 456 items, 139 (30.5%) appeared after he was 70. While he continued to publish on fossil amphibians and reptiles (the last five articles on these groups appearing in 1950), on the wealth of non-australopithecine fossils found in various cave-filling breccias in the Transvaal, and on various other topics (tumble-weeds, premolars of elephant shrews, postage stamps, organs of Jacobson in pangolins, etc.), he still published 68 articles on australopithecines and other aspects of human evolution after he was 70. The last one, an important monograph on robust australopithecines, appeared in 1952, a year after his death.

Broom spent the major part of the first half of 1937 in the United States, a country he had visited and worked in so often (usually at the American Museum) that he regarded it as a second home. He lectured across the country from New York to Berkeley, saw Hollywood, swam in Great Salt Lake, talked to pretty girls, received an honorary doctorate from Columbia University, and in general had a triumphal tour.

Things had languished at Sterkfontein in Broom's absence, but picked up immediately upon his return, with the retrieval of many specimens. On these, in addition to doing much of the collecting, he did most of the tedious laboratory work himself, made the drawings, and even made the molds for casts, while still studying and publishing on his amphibians and reptiles from the Permian of the Karroo. In June of 1938, unexpectedly and indirectly, he discovered a new site and a new kind of 'ape-man' (his term, not mine).

The full story has been told better by Broom himself (1950:49-51), but can be abstracted here: At Sterkfontein, Barlow had given Broom an australopithecine palate with an M^1. Broom's keen eye recognized what Barlow seemingly did not; the shapes of the palate and of the tooth were different than those of other specimens from Sterkfontein and the matrix adhering to the bone was also different. Broom bought the specimen, saying nothing, but did some detective work; talking to the workmen while Barlow was absent, he learned that none of them had seen the piece. A few days later he faced Barlow again, whose original reticence had not been explained, and got from him the truth; the specimen had been brought from somewhere else by a schoolboy of 15, Gert Terblanche, who served at Sterkfontein as a tourists' guide on Saturdays. Broom tracked the boy to his country school, retrieved four teeth of the jaw from his pocket, lectured on fossils and cave-fillings to the whole school, and pursuaded Gert to take him to the spot of original discovery. There was found a partial skull which Gert had broken badly trying to extract it from the rock. Thus was Kromdraai discovered; Broom collected all the fragments, reconstructed them, and had the type specimen of the first-recognized robust australopithecine, which he called _Paranthropus robustus_ (Broom, 1938a), a name still sometimes used.

Due in part to stimulus from Broom while in America, and then to direct invitations from Broom and Dart, Wm. King Gregory and his colleague, Milo Hellman, visited South Africa in 1938, studying in particular the dentitions of all the australopithecines (a term not in use at the time) collected to that date. Their descriptions and conclusions (Gregory and Hellman 1938; 1939a, b,c,;1940) marked the beginning (but only a beginning) of the scientific world's re-assessment of the place of _Australopithecus_ in human evolution.

The essence of their conclusions, as Gregory (1930) had already stated several years earlier when he had first seen only photographs of the teeth of _Australopithecus africanus_, was that the individual teeth and the assembled dentitions were much more like those of humans than like those of any known apes, that the teeth could easily have evolved from those of dryopithecines, and that they could logically have been ancestral to the teeth of 'man' (i.e., *Homo*, *Pithecanthropus*, and *Sinanthropus*, as then understood). Gregory and Hellman contradicted Abel's con-conclusion of 1931 that the dentition of *Australopithecus* was similar to that of a gorilla, and additionally created a new subfamily, the Australopithecinae (1939b) for the gracile and robust australopithecines as then known (*Australopithecus*, *Plesianthropus*, and *Paranthropus*).

They did not, as is sometimes stated, place this new subfamily in the Hominidae; instead they rather curiously presented formal definitions of the Australopithecinae and of the Homininae, without assigning either subfamily to a family. The Homininae obviously belong to the Hominidae, but if no other subfamily is to be assigned to that family, why designate a subfamily? If the Australopithecinae, as defined, were regarded as apes, why were Gregory and Hellman so indefinite? True, in the text of the article the australopithecines were called apes: "...we affirm with Broom that in South Africa there once lived apes which had almost become men.", and again they referred to the australopithecines as "...conservative cousins of the contempory human branch."[19] This lack of a definite assignment to a family is consistent with all of the other writings on the australopithecines to the end of the fourth decade of the Twentieth Century; except for Adloff (1932), one does not find formal taxonomic terms used, which lack of systematic clarity is probably one reason for much of the taxonomic and intellectual confusion throughout this early period.[20] Habits once formed are difficult to abandon; thirty years later, Dart (1967, p. 135) still was calling australopithecines "man apes" and pithecanthropines "ape men".

Broom was not one to let anyone else, even his guests, get ahead of him; in the same year and the same volume (Broom 1939a) as Gregory and Hellman's major publication (1939b), he also published a description of the dentitions of the known australopithecines, and reached their same conclusions, which of course he already had. Two years later, Senyurek (1941) published his study of the casts of only 14 australopithecine teeth, but reached the very similar conclusions that the teeth were more like those of humans than of apes.

In spite of this concurrence of opinion, and particularly the strong support of Gregory and Hellman, Broom continued to use the words 'ape' or 'ape-men' in referring to the australopithecines. Additionally, having pulled back from his published opinion of 1930 that the australopithecines were Pliocene in age, he consistently (Broom 1939b) in the late 1930s and into the 1940s held that they were Pleistocene in age, whereas by contrast "men" (he obviously had Piltdown in mind) were already present in the Pliocene. Thus australopithecines could not have been ancestral to 'men' but were instead relatively unchanged descendants of those ancestors.

Another pattern of thought in Broom's mind, as more broken skulls of gracile and robust australopithecines appeared from Sterkfontein and Kromdraai, was the problem of size of australopithecine brains. Dart had estimated the endocast of the original type of *Australopithecus africanus* to have been 520 cc., larger than that of any known chimpanzee but less than those of some gorillas. Broom, however, tended to let his estimates grow, particularly where the large-jawed *Paranthropus* was involved, and sometimes published fanciful figures of more than 800 cc., based upon no evidence whatsoever. Subsequent research has reduced the range of volumes of endocasts of South African australopithecines to 435-540 cc (mean = 494 cc) for 6 gracile specimens (Tobias 1971, p. 20). Other investigators, as discussed by Tobias, have published slightly lower ranges and means. The size of endocasts of robust specimens falls within the same range as that for the graciles.

As a signal of greater things to come, but a signal unknown at the time, the first fossil of an australopithecine to be found outside of South Africa was collected in northern Tanganyika (now Tanzania) early in 1939. The specimen consisted of a small fragment of a maxilla with p^{3-4}, collected by Dr. F. Kohl-Larsen, director of a German ethnological expedition, the members of which seem also to have picked up numerous fossils. Due probably to the stresses induced by the Second World War, these two teeth (plus an isolated upper molar which may or may not have belonged to the same population) were not described and named until more than a decade later (Wienert 1950), with the teeth being described in greater detail the following year (Remane 1951). Wienert assigned the fossil to *Meganthropus*, as *M. africanus* sp. n. This genus had been published by Weidenreich (1944) for a fragment of a massive lower hominid jaw with the premolars and first molar, collected in Java by G.W.R. von Koenigswald in 1941. Von Koenigswald had sent a cast to Weidenreich, with the suggestion of the name *Meganthropus paleojavanicus* gen. et sp. n., but, with the coming of war to Indonesia, von Koenigswald had disappeared into a concentration camp from which there was no communication. After waiting several years, Weidenreich with some reluctance published the description and new name as based on the cast he had, not knowing what had happened to either the original specimen or its discoverer.

The assignment in 1950 of the two upper premolars collected in Tanganyika in 1939 to a genus represented only by a piece of lower jaw from Java was not logical, since no comparisons could be made, and in 1953 John Robinson published his conclusions that the specimen upon which *Meganthropus paleojavanicus* had been founded was more probably a Javanese representative of *Paranthropus*, whereas *M. africanus* most likely was merely an eastern African member of *Plesianthropus transvaalensis*. The latter decision was seemingly a sensible one and has been generally followed, although Remaine (1954) objected that the two premolars of '*M. africanus*' were much too primitive in their morphology to be considered australopithecine. The extension of range of the robust australopithecines to Java, however, has not met with equal acceptance, and as subsequent years have passed without further evidence of australopithecines outside of Africa the possibility of having robust individuals in far Asia has seemed less reasonable.

During the years of the Second World War (1939-1945), paleo-anthropological collecting slowed for Broom, but other activities did not. Sterkfontein was closed as a commercial lime-site, and, say what one will about blasting as a method of exposing fossils, without money of one's own Broom found no other way to open large areas. The Museum's budget was cut and its truck sold; many of the staff had gone to war, gasoline was unavailable, and publishing was curtailed for the duration. Broom continued research on specimens he had in the Museum and wrote some articles on other topics; he found other sources for his articles, so his flow of publications was hardly slowed. Additionally he was working on a monograph on australopithecines. However, a gap of six years of active field-collecting must have irked a man as active physically and mentally as Broom, particularly when most of that six years occurred after he was 73, but no complaint occurred in his little book, "Finding the Missing Link" (Broom 1950). He did considerable local lecturing, and carried on much correspondence with Arthur Keith, D'Arcy Thompson, and W. E. Le Gros Clark.

The one exception to the lack of fieldwork was a small bit of quarrying in 1941 at the spot at Kromdraai where Gert Terblanche had found the type specimen of _Paranthropus_. Although mentioned earlier, the name of John T. Robinson enters our chronologic history here for the first time as a new assistant in charge of the work. A remarkable collection was made: a lower jaw of a three-year old with a nearly complete and unworn deciduous dentition plus intact crown of the first molar, distal end of a humerus, proximal end of an ulna, part of a hand, and talus (astragalus, one of the bones of the ankle). The pieces of humerus and ulna were so human-like that Broom wrote (1950, p. 55) ". . . practically every anatomist in the world would say they were undoubtedly human." The remnants of the hand were sufficient to show that its owner was not quadrupedal, as Keith had suggested for _Australopithecus africanus_ in 1931, and Broom's study of the talus convinced him that its possessor (probably the same individual to whom the hand belonged) was undoubtedly bipedal (Broom 1943). Evidence, at last, was beginning to accumulate that Dart's conclusions of 1925, that his original specimen from Taung represented a bipedal population, was correct.

However, in spite of such evidence and in spite of the close similarity of the dentitions of the australopithecines to those of living men and the latters' immediate pithecanthropine ancestors (Gregory and Hellman, 1939b, c) most people remained unconvinced of any validity of a close relationship of the australopithecines to 'man' (i.e., themselves), but seemingly assumed instead the past presence of several kinds of strange apes which had become extinct without issue. People remembered that Dart had called his original specimen an 'ape.' As mentioned before, Broom himself fostered this view by continuing to call specimens of australopithecines, or the group as a whole, either 'apes' or 'ape-men.' Gregory and Hellman (1939b, c) used the term 'man-apes,' but always the word 'ape' was present, and Adloff's clear statement (1932) that *Australopithecus africanus* was hominid was completely disregarded in practice for many years thereafter.

Thus in 1945 both G.G. Simpson's "The Principles of Classification and a Classification of Mammals" and the second edition of A.S. Romer's "Vertebrate Paleontology" listed the australopithecines as belonging to the Simiidae (=Pongidae). Even the scientific community was not then ready for early Pleistocene or late Pliocene hominids that didn't yet look like they thought hominids should; not only was the large-brained, ape-jawed Piltdown specimen a model of what 'early man' should be, but many physical anthropologists of the time expected hominid ancestors as primitive as the australopithecines were to have lived much earlier, since (as also mentioned before) they considered the division between hominids and other primates to have occurred in the Oligocene or earlier, (thus avoiding relationship with apes), a view which has lingered on (Genet-Varcin 1969; Kurten 1972).

In 1945, several months before the end of World War II, a group of Dart's students visited the valley of Makapan,[21] in northern Transvaal, stimulated by the knowledge that many caves existed there, that mammalian fossils had been sent from some of these caves to Dart in 1925, and that archeological materials of both the Middle and Old Stone Ages had subsequently been recovered from some of the caves. In addition to the known sites, the group visited a commercial limeworks further down the valley, and there found many mammalian bones in breccias on the discard-dumps. Among these was the skull of a fossil baboon, *Parapapio broomi*, known from Sterkfontein and so associated there with gracile australopithecines. The leading spirit among the students was Philip Tobias, who took the baboon's skull back to Dart and enticed him to renew his activities in paleo-anthropology. Work was begun in September 1945, and continued intermittently, as money and manpower were available, for several years.

In the meantime, Broom had finished the manuscript of his part of a monograph on all australopithecines found before 1942. The second part of the same monograph was written by G.W.H. Schepers, who had been one of the medical students who had first led Broom to Sterkfontein in 1936; in the meantime, Schepers had finished medical school with a specialty in neurology, had taught in Dart's department for several years, had acquired a Ph.D. and a D.Sc. from the University of South Africa, and had become Professor of Anatomy and head of the department in the medical school of the University of Pretoria. Broom's part of the monograph, the longer by far, was on dentition and osteology, while Schepers contributed the part on the endocasts, of which five examples were by then available. The manuscript, in a near-finished state in 1944, was shown to Field Marshall (and also at that time, Prime Minister) Smuts, whose enthusiasm for science was matched by his political power; the Prime Minister stimulated the formation of a National Research Council, which then underwrote the costs of publication as one of its earlier acts. The Prime Minister wrote a Preface, and the volume was published in January 1946 by the Transvaal Museum.

The book received rave reviews from many (the U.S. National Academy of Sciences declared it the most important book of the year in biology), and Arthur Keith, who as we have noted had been most resistent to the idea of hominid affinities for the australopithecines, wrote Broom that the monograph had changed his mind. A year later Keith (1947) published his retraction, but (like Broom and others) could not bring himself, not quite, to call the australopithecines hominids: "...ground-living anthropoids, human in posture, gait and dentition, but still anthropoid in facial physiognomy and in size of brain." In the same short article he introduced the informal name of 'Dartians' for the australopithecines, claiming the latter name was too long and clumsy; this substitution in general has not been used.

Other reviewers, however, had reservations, on several grounds. The monograph lacked the measurements, except for some of the teeth, which were generally common in paleontological publications and indeed were valuable in allowing specialists who could not study the original specimens to evaluate their size and proportions. The text-figures were Broom's typical quick pen-and-ink sketches, which some critics thought could be more interpretive and thus less desirable than the accuracy of a photograph; indeed, of the 13 plates, only two were photographs, the others consisted of drawings.

The severest criticisms, however, were of the interpretations by Schepers of the several partial endocasts, natural and prepared, which he had studied. In large part Schepers disregarded the known evidence that in modern humans the width of the fluid-filled cavity between brain and the bone enclosing of the cranial cavity is such that an endocast produces an accurate record of the inner configuration of the skull but little of the detail of the brain; indeed, brains taken from fresh human cadavers do not, in pattern of sulci and gyri, correspond to impressions of the endocasts made from the same skulls (Symington, 1916). With particular regard to endocasts of fossil hominids, Le Gros Clark (1938) had warned of the difficulties and dangers of identifying sulci and gyri from endocrancial casts. Schepers did not mention these publications or problems in his portion of the volume, although he did mention that considerable skepticism did exist among paleo-neurologists about the value of endocasts of anthropoid fossils.

Most of the adverse criticism of Schepers was because, in the opinion of other anatomists, he read more into the endocasts than was possible to know, extrapolating from difficult or indefinite anatomical identifications to rather precise neurological functions and individual and social behavior. Particularly his claim for the capacity of speech in the australopithecines was received with skepticism. Additionally, Schepers' personal view of evolutionary theory and the evolution of Primates in particular, with separation of ancestral hominids in the Eocene (his text) or the early Oligocene (fig. 33), did not find as much favor at the time as it would have earlier, and has been increasingly ignored since. Further, other neuro-paleontologists who have looked at his casts or made new ones for their own studies cannot find all of the detail which Schepers reported, and believe that he allowed more freedom for his imagination than was wise. Even before these latter criticisms could be made, however, the negative reactions against particular aspects of Schepers' section of the monograph tended to be reflected upon the work as a whole.

One might think that an endocast, natural or prepared, of the cranial cavity of an australopithecine might well preserve more detail of the surface of the contained brain than would be true for a modern human, with a brain nearly three times the size, inasmuch as a long series of publications by Leonard Radinsky at the University of Chicago has shown a close identity between outer surfaces of brains and inner surfaces of adjacent cranial bones in many small to medium-sized mammals. Only in the larger mammals do the arachnoid space and the dura mater separate brain from bone so thoroughly that, as in living humans (Symington 1916), the pattern of the one does not reflect in any detail the pattern of the other. However, seemingly the chimpanzee, with an endocranial cavity almost as large as that of a typical australopithecine, is such a 'larger mammal,' and little of the detail of the surface of the brain is impressed upon the adjacent bone of the skull (Le Gros Clark et al. 1936).[22] While this latter research should be repeated by appropriate studies of gorillas, orangutans, and other chimpanzees, the possibility seems remote that we shall ever be able to ascertain the wealth of data from the study of australopithecine endocasts which Schepers thought he could.

Our story now divides into three intertwined parts; one, the continuation of Broom's own research and thought, naturally ended with his death in 1951 (or, if one prefers, with the publication of his last monograph in 1952). A second line of endeavor is that of the work of Dart and an important group of colleagues (Alun R. Hughes, Philip Tobias, James W. Kitching, C.K. Brain and others) at the Limeworks Site in the Makapan valley, northern Transvaal. The third is the introduction of statistical techniques into the study of australopithecines, a pattern of thought and action precipitated by Solly Zuckerman when he was adversely affected by the conversion of Le Gros Clark to Dart's and Broom's concepts of the close relationship of the australopicines to humans. The second and third of these lines of thought and research continue beyond the time of the death of Broom, into the present, and so here will be traced only in part and only in outline.

Even after the Second World War was finished, Broom's collecting activities were delayed almost a year and a half, due to the formation of a Historical Monuments Commission which, among other activities, was empowered to have jurisdiction over all collecting of fossils in South Africa. Trying to control Broom was like trying to harness the whirlwind, and the South African press, which always found Broom to be cooperative and news-worthy, made the Commission look foolish, as in the instance when the latter laid down the rule that Broom could not collect unless accompanied by and cooperating with a "competent field geologist." Broom, who had been a medallist in geology as an undergraduate, Professor of Geology 1903-1909 at the University of Stellenbosch, and had studied as much of South Africa's sedimentary geology as almost any man alive, was naturally outraged, and planned to break the Commission's rule deliberately, hoping to be arrested so that he might make the case into a cause celebre. His own Museum's Board wouldn't let him, and he owed too much allegiance to the Pretoria Museum (and had too many of his fossils there) to break with them.

On the 30 November, 1946, Broom had his eightieth birthday. Honors flowed in from all over the world, and the Royal Society of South Africa planned a commemorative volume for him, which was issued two years later (Du Toit 1948). This volume, with chapters written by different specialists from around the world, emphasized Broom's research on Synapsida and Australopithecinae, but most of the remaining riches of his life were hardly mentioned; the addition of articles on all the subjects of Broom's contributions would have been impossible within the limits of a single volume.

Probably the most important happening of 1946 for the history of the study of australopithecines was the visit to South Africa of Wilfred E. Le Gros Clark, who was at that time Professor of Anatomy at Oxford. Born in 1895 (and thus two years younger than Dart), he was appointed Professor of Anatomy at the University of London at the age of 29, and served there for ten years before moving to Oxford in 1934. He was a keen student of the morphology and evolution of Primates, an authority on the structure of tree-shrews and some of the insectivores, the author of a text on tissues, and he had done experimental research on regeneration of mammalian muscle. By 1946 he had written one book (1934) on the evolution of Primates, and was destined to write several more. He was a worthy successor to the group of British anatomists and paleo-anthropologists under whom he and Dart had trained in the early 1920s, but unlike them he decided--stimulated by the monograph of Broom and Schepers--to go to South Africa and see all of the original australopithecine specimens himself. He had just published an article on problems of human paleontology (Le Gros Clark 1946) and then prepared himself further by studying, and taking notes on, the skulls of more than a hundred apes in British collections. He arrived in Johannesburg late in 1946, spent several days studying Dart's original specimen of *Australopithecus africanus*, and then went on to Pretoria for almost two weeks of intensive study of Broom's collection of what were then called _Plesianthropus_ and _Paranthropus_. He had arrived extremely doubtful of the validity of the claims of Dart and Broom as to the similarities of the australopithecines to 'man'; he departed convinced, in large part, of the validity of those conclusions (Le Gros Clark 1967, p. 32).

Broom, Le Gros Clark, Dart and others from South Africa then went to the First Pan-African Congress on Prehistory, organized by L.S.B. Leakey[24] and held in Nairobi, Kenya, January 14-23, 1947. The Abbe Breuil was elected president and Broom became vice-president, but more important, Le Gros Clark threw the weight of his learning and influence to support the concept of the hominid-like nature of the australopithecines; indeed, almost half of the report on the Congress appearing in NATURE some three weeks later (Anonymous 1947) was devoted to Le Gros Clark's views. For a change, in a scientific report on australopithecines, the term 'Hominidae' was used in the sense of formal systematics, instead of the vague term 'man.'

Le Gros Clark then returned to England, with the zeal of a crusader for the cause of hominid affinities of australopithecines; he lectured and wrote, getting two major articles published that same year (Le Gros Clark 1947a, b). In so doing, he met a strong opponent in Solly Zuckerman, a story which will be followed in part later in this chapter.

Early in January 1947, before leaving for Kenya, Broom had resumed his field-research, with the assistance of John Robinson and several well-trained and highly competent Zulus. After Broom returned from Kenya near the end of the month, they worked variously at Kromdraai and Sterkfontein; in the absence of commercial activity, Broom and Robinson undertook their own dynamiting, carefully planned and controlled. Under this regime, the results that year were spectacular; the finds were too numerous to list here (see Broom, 1950, pp. 63-72). The most important finds came from Sterkfontein; of these two have become particularly famous: 'Mrs. Ples' (Sts 5), the most complete and perfect skull of an australopithecine found to that time (Broom 1947) and Sts 14, the major part of a pelvis, with part of the proximal femur, two lumbar vertebrae, and some pieces of ribs, all undoubtedly from the same individual (Broom and Robinson, 1947).

The remains of Sts 14, a dainty female with a height of ca. 130.6 cm. (=4 ft. 3 in.) and an estimated weight between 18.2--27.3 kg (=40--60 lb., with a figure between 50 and 60 lb. thought to be more probable by myself) have been studied by many, but particularly by Robinson (1972). The important aspect of the pelvis, as seen immediately by Broom and Robinson when it was first found, was the greater similarity, particularly of the broad, short ilium, to that of living man and the considerable dissimilarity to that of any known ape. This similarity to modern man seemingly presented valid evidence for bipedal locomotion, at least for gracile australopithecines, but even so the critics beyond (and some in) South Africa were for the most part not convinced thereby that the australopithecines were not apes, or in some cases they were not convinced that they had bipedal locomotion or if that was true the animals must have waddled as they walked, or they couldn't walk but perhaps could run. More recent studies indicate that there was little justification for most of these statements; seemingly, with bipedalism as with all other characteristics of the australopithecines, there existed among most paleo-anthropologists an unconscious desire to proclaim the australopithecines to be less human, and thus more removed from self, than most of us presently believe to have been true.

The work of Broom and his helpers continued through 1947 and into 1948, mostly at Sterkfontein. Increasingly Broom depended upon Robinson, both for the success of the field-work and in part as co-author on the numerous publications (mostly appearing in NATURE) on the australopithecines, but occasionally also on fossil reptiles from the Karroo. This cooperation continued over the next few years, and by November, 1950, Broom could write to Le Gros Clark, "I am not so necessary now. Mr. Robinson is as good as I am." But that was yet in the future.

In September, 1948, Broom teamed up with a young American, Wendell Phillips, who had more money (not his own) and egotism than scientific ability. Broom, almost 82 at the time, seemingly was taken in, in part at least, by the glib young Phillips, then less than 30, who had organized a double-pronged expedition for the scientific exploration of Africa, by talking the U.S. military (mostly as represented by Admiral Nimitz) out of equipment and gasoline;[25] establishing a pattern which was soon to lead him to becoming a multi-millionaire (Anonymous 1966), he talked numerous people out of considerable sums of money, all for the sake of science.

Phillips led his part of the expedition into Africa at Egypt and worked his way south,[26] but Dr. Charles Camp, vertebrate paleontologist at the University of California (where Phillips had studied as an undergraduate) and the director of the southern half of the expedition, had separated himself from Phillips by the time they both arrived in South Africa. Broom allied himself with Phillips, perhaps because of the available money, while Camp received the blessing of, and the permits for excavation from, the Historical Monuments Commission.

Camp's team of scientists was large, versatile, and well organized, covering studies from geology through paleontology into most phases of natural history and on to human ecology, ethnology, linguistics, and musicology (Camp, 1948). With regards to australopithecines, they made a survey of the Makapan Valley, where Dart and his crew were working (Barbour 1949a, b,) and carried this type of study south to the area of Taung, where Frank Peabody made a careful study of the geology and associated prehistoric fauna of the area from which the type specimen of Australopithecus africanus had been taken almost a quarter century earlier. This study (Peabody, 1954) for almost another quarter-century was the primary source of knowledge for geology and prehistoric environment and fauna of the area of Taung. The other contribution of Camp's crew to australopithecine studies was the opening during the spring and summer of 1948 of numerous small sites (Bolt's Farm and others) near Sterkfontein. Camp (1948) mentioned finding australopithecine femora, which must have irked Broom, and Camp's collection of numerous mammal-like reptiles from the Karroo must have irked him more; in any case, Broom said some unkind things about Camp and his people, for which Smuts asked him to apologize but no evidence exists that he did so.

The important aspect of the cooperation between Broom and Phillips was the opening of a new australopithecine site, that of Swartkrans, in September, 1948. This site, only about a mile from Sterkfontein, produced numerous examples of the robust australopithecines, which Broom regarded as representing a new species, Paranthropus crassidens (Broom 1949a). The work at Swartkrans proceeded through 1950, yielding a wealth of broken bones of australopithecines, some of it post-cranial. Among the post-cranial remains was a partial innominate (Broom and Robinson 1950a) which had a broad, short ilium similar to that of Pleisanthropus and to that found in 1949 by Dart at Makapan (Dart 1949 a, b), but the remainder of the pelvic fragment was enigmatic. Still, the ilium indicated that the robust australopithecines, quite like the gracile ones, must have been bipedal. Another unique find was a lower jaw with such a low ascending ramus, in contrast to the large high rami of robust australopithecines, and with molars the size of _Pithecanthropus_ in contrast to the huge molars of the robust forms, that Broom and Robinson (1949) considered it to be a non-australopithecine, and thus a 'true man.' They named it _Telanthropus capensis_ gen. et sp. n. This jaw was found in a pocket of the main breccia which had been eroded out and then refilled, with the jaw in the new filling; the status of the jaw is still being argued, some claiming that its proportions could fall within the normal range of variability of the australopithecines, but others believe that conclusion impossible and, thinking too of the presumed later period of time, consider _Telanthropus_ to have been a southern representative of a post-australopithecine population of _Homo_.

These post-war years, with excavations at Sterkfontein, Kromdraai, and Swartkrans, was a period of intense activity. Broom's scientific production was remarkable: 13 publications in 1946, 14 in 1947, 8 in 1948, 18 in 1949, and 16 in 1950. He did not narrow his interests; although the australopithecines and the fossils of mammal-like reptiles from the Karroo occupied his main attention, during these years he also worked and published on golden moles, rare postage stamps, fossil baboons, modern orang-utans, a gorilla's foetus, the first white rhinoceros in captivity, flight in ancestral ostriches, the skull of the American horned lizard, the extinct blue-buck, deciduous dentitions of the living large felines, some thoughts on religion and evolution, and memories of the chemistry laboratory at the University of Glasgow 60 years earlier. During this time he also found time to prepare the manuscript for a charming little book, "Finding the Missing Link," (Broom 1950) and for an intensive monographic study of the australopithecine fossils from Sterkfontein (Broom and Robinson 1950; Schepers 1950), and to begin the study, writing and drawings for yet another monograph, this one on the fossils of _Paranthropus_ from Swartkrans, which he finished before he died.

Broom was accustomed to working at such a rapid pace, and might have continued for several years more, in spite of some signs of failing health, if he had not, early in 1949, accepted an invitation to visit London to receive the Wollaston Medal, the highest award of the Geological Society (British). The medal was to be given in late April, but Broom went a month earlier to the United States, where he lectured from coast to coast and appeared at numerous receptions, banquets, press conferences, and on television shows. He had very little sleep and was traveling constantly. In England he had much the same routine, visited and lectured in Scotland, and even lectured in Utrecht, The Netherlands. In spite of the resultant exhaustion, his acumen was not diminished, as witnessed by the fact that in the Hunterian Collection of the University of Glasgow, Scotland, he suddenly recognized a skull of a large antelope as being different than any known before; he was able to deduce that the skull was from a South African blue buck (Hippotragus leucophaeus) extinct since ca. 1799. No skull of this population had hitherto been known; the skull had been lying unrecognized in the collection for at least 150 years. Broom published on this incident, of course (1950).[27]

Broom returned home more exhausted than he knew, or would admit; he plunged into work again but soon collapsed. Sickly as a child, Broom had suffered from asthma through much of his earlier years, but had finally cured himself in part by learning to avoid eggs in his diet. The asthma now returned and with it a weakened heart; he had spells of dizziness, an irregular pulse, and difficulty not only in breathing but in swallowing and speaking, particularly after any strain or fatigue. Under the best of circumstances, he could work at the museum only a few hours at a time, but often he had to stay home, and sometimes he was convinced he was dying. If anything kept him alive during that winter of 1949 (summer in the northern hemisphere) it must have been willpower, because he and Robinson had embarked on a monographic study of the australopithecines from Swartkrans, and the old man was determined to see it through. Broom was essentially religious in a fashion uniquely his own, and gave full credit to the Diety for his gradual, slow, and partial recovery.

During 1950 Broom was better; he worked at the museum oftener and longer, but even during his bad periods he would prop himself up in bed and, specimen in hand, write or draw. He followed the battle between Le Gros Clark and Zuckerman, being waged in the English journals of the time over the affinities of the australopithecines, with great pleasure, knowing full well in his own mind that Zuckerman, in his use of statistics, could not find anything that an experienced morphologist with a good eye would not have already seen.

In Broom's mind, Zuckerman was not a morphologist, had a most undiscerning eye, and was simply wrong in his rejection of the hominid affinities of the australopithecines; thus his recourse to statistics, which Broom simply didn't understand, was considered by the latter to be an exercise in futility.

Toward the end of 1950, Broom's health worsened, and during January and February of the next year he was near death at least twice, but each time recovered enough to continue work on the memoir on Paranthropus from Kromdraai. For inclusion in the monograph Broom had prepared a reply to Zuckerman, two of whose main mensurational and statistical papers had been published the year before (Ashton and Zuckerman 1950 a, b), but he could not really state anything from his own morphological point of view that had not been published before either by himself or others, and the tone of his reply has been reported as being definitely polemical, so wiser heads urged him to delete the material. He therefore worked on pruning that part of the manuscript through much of March 1951. The finished result, as published a year later (Broom and Robinson 1952), appears on p. 31 and pp. 99-104. Each reader will have to judge for himself; I find the criticisms by Broom (and Robinson?) to be just and carefully considered. The polemics, such as may have existed, have been removed.

Early in April 1951, Broom worked at the museum for the last time, completing the last drawing for the monograph, but his strength was failing, and he was soon confined to bed under nursing care. Mary Broom herself was near total exhaustion with her constant care of him for the preceding several months. On the morning of April 6, Broom sat up and asked for pencil and paper. He wrote some business letters, wrote some checks, and made the final additions and corrections to the monograph. A nephew then heard him mutter, "Now that's finished. . .and so am I." He then whispered to his wife, "I am very ill, darling," and according to her did not speak again; paralysis had set in and he could not. He then died that evening.

The monograph on the robust anstralopithecines from Swartkrans, upon which he had worked his last day, was published eleven months later (Broom and Robinson 1952).

During June of the year before Broom died, a meeting of naturalists, anthropologists, and human geneticists was held at Cold Spring Harbor, Long Island, New York, dedicated to the topic of the origin and evolution of man. One trend that emerged during the course of the meeting was the acceptance of australopithecines as hominids. This trend, minor at the time, was in later years to grow to become the dominant one, in spite of the vigorous opposition of Zuckerman and his colleagues, an opposition which--if measured by number of publications--developed dramatically between 1950 and 1956. Strangely, for a meeting devoted to human evolution, no European or South African paleoanthropologists attended, although several European geneticists were present. With the exception of Le Gros Clark, Europeans of that time, and even Dart, Broom, and other South Africans, would have classified the australopithecines as pongids, but at that meeting at Cold Spring Harbor in June, 1950, Sherwood Washburn (1950), William W. Howells (1950), and Ernst Mayr (1950) placed them firmly in the Hominidae.[28]

Ernst Mayr, ornithologist and general student of evolution,[29] went much further in his conclusions. First he stated that the single genus of _Drosophila_ (fruit-flys) had as much, or almost as much, diversity among its more than 600 species as existed in the whole order of Primates, or at least as much as in the sub-order Anthropoidea.[30] Following that example, thus, all Primates might be classified as _Homo_, or at least all of the Anthropoidea should be classified as _Homo_. However, realizing how grotesque such a recommendation would have been to most anthropologists, and how ridiculous it would seem to the public, he modified his suggestion to include both the australopithecines and the pithecanthropines in _Homo_, which genus would then be considered to have only three species.

Both robust and gracile australopithecines were classified as one species, _Homo transvaalensis_,[31] on the basis that insufficient knowledge at the time precluded any other decision; all of the pithecanthropines were put together into _Homo erectus_, a recommendation made earlier and independently by Mayr (1944) and Evans (1945).[32] All populations more recent than those admitted to _H. erectus_ were classified as _Homo sapiens_. This scheme of only one genus with three species for the Hominidae (_Ramapithecus_ not then being considered a hominid) was then and mostly since has been too extreme for most anthropologists to accept, although the inclusion of the pithecanthropines within a single species of _Homo_ has become widely accepted. However, the generic name *Australopithecus* has, by general practice, held its place, but usually with at least two species, *africanus* and *robustus*[33] for gracile and robust australopithecines, respectively. The proposal by Robinson (1967, 1972) to include the gracile populations in *Homo* while maintaining the robust ones in *Paranthropus* has been little followed, although if the gracile australopithecines are ancestral to all later hominids Robinson's taxonomic pattern is a logical one.

We do not know that Broom was ever aware of the support received from Washburn and Howells, or of the taxonomic revision proposed by Mayr; if he knew of these matters he did not mention them in his last publications.

We now move back a few years to pick up another thread of our story. When Le Gros Clark was working in Dart's laboratory in Johannesburg late in 1946, he made a remark to the latter which set Dart off onto a new kind of research which occupied much of his energies for many years and produced the concept of the 'osteodontokeratic culture.' The two men were talking about Dart's observations of many years that the skulls of fossil baboons seemed often to have double depressed fractures, as if the animals had been killed by being hit on the head with a double-headed club. Le Gros Clark suggested that Dart keep a tally of such cases as contrasted with those seemingly not so killed. This led Dart eventually to analyzing and counting all of the fragments of all of the bones which he and his coworkers excavated from Makapan for the next several years.

Extracting tens of thousands of bits of broken bone from hard breccia is slow, tedious, and difficult work, and expensive too, and could not have been accomplished had not Bernard Price, a Scots electrical engineer who had made a fortune in South Africa, come to the rescue financially by founding and supporting the Bernard Price Institute for Paleontological Research, which continues to function at the University of Witwatersrand. Beginning in March 1947, with assured financial support, Dart maintained a permanent crew of a few experienced men in the Makapan Valley, concentrating first at the Cave of Hearths. In their spare time, however, one or more of the men explored the bone-filled breccias from the mine-dumps at the 'Limeworks,' where Tobias had found the skull of a baboon which had stimulated Dart to resume his paleoanthropological studies. From one such piece of breccia, James Kitching in September 1947 extracted an occipital bone which Dart recognized immediately as being from an australopithecine (Dart 1948). The lack of massive nuchal muscles, the low position of their attachment on the rounded occiput, the poise of the skull on the vertebral column, and several other characters discussed by Dart proved to him the much closer affinities of his new Australopithecus prometheus (as he named the species)[34] to hominids than to apes.

Thereafter the work of recovery of fossils intensified at the 'Limeworks' site, with new and important specimens being extracted almost continuously, quite as Broom had done in the late 1930s and was doing even then at Sterkfontein, Kromdraai and Swartkrans. Broom, with each new major discovery, typically prepared a short article for NATURE: Dart tended to use the AMERICAN JOURNAL OF PHYSICAL ANTHROPOLOGY in the same way. Various pieces of cranium and jaws kept appearing, and these were quickly recorded, drawn, photographed, discussed, and published. However, these remains of australopithecines were only 0.43% of the total of the osseous fragments, the remainder being bones mostly of antelopes but many other kinds of mammals were also represented, among which baboons were numerous. During this period of the late 1940s, Dart was developing his ideas as to why the bones of all these mammals were represented in the cave-residues of a site which he visualized as having been the living-quarters of australopithecines.

Neither Dart nor seemingly anyone else during these early decades of investigations on australopithecines considered the several 'caves' as other than walk-in, live-in sites which the australopithecines had used for occupation, either in family groups or in larger social organizations.[35] This concept was a natural one, considering that at the time most of the best-known concentrations of human fossils and/or prehistoric human artifacts came from caves of Cro-Magnons, Neandertals, or pithecanthropines in Europe, southwestern Asia, or northeastern China, all sites well-known to South African prehistorians. One has only to look at the pictures of the australopithecines in F. Clark Howell's "Early Man," (1965, Second Edition 1973) to realize the persistence of this concept that australopithecines lived in caves, carried parts of slaughtered or scavenged animals to their caves, and later sat in their caves and there fashioned the crudest of bone tools--so crude that most people do not recognize them as tools.

The simple fact is that the Transvaal is full of limestone caves, and that some of these have secondarily been filled with lime-breccias which often contain thousands of broken bones of mammals, including a low proportion of australopithecines. Dart's consideration of these factors, and his further studies of the numbers and kinds of the broken bones from the several sites being excavated, but particularly at Makapan, led him to several conclusions, which, beginning in the late 1940s, he published in one form or another at every opportunity and in which he continues to believe firmly.

These conclusions can be summarized as follows: the australopithecines used the caves as habitation-sites; the gracile australopithecines, at least, were meat-eaters, and they procured their meat by the hunting primarily of small to medium-sized animals; the animals were often if not usually killed by being clubbed or stabbed; the tools so used for clubbing or stabbing were the larger limb-bones and/or the horns of the larger ungulates (salvaged on the veldt from the kills of hyenas, lions, or leopards); jaws of the larger carnivores and of porcupines were also purposely collected for intended use of the specialized carnivores' canines and the rodents' incisors; all bones collected were carried to and into the caves being inhabited, and there were purposely broken in definite patterns, and then often broken or chipped further to produce a variety of recognizable shapes which were, indeed, tools; the tooth-rows of ungulates were salvaged for use as rasps particularly; certain other bones of certain other animals were also collected for particular specialized purposes. These ideas were developed over several years from the study of fossils from Makapan and supported by Dart's careful census, recorded by species and bones of each species. Remains of some animals were more numerous than others; remains of some bones of some animals were more numerous, and even some parts of the broken bones were more numerous than other parts of the same bones. To Dart all of these data had meaning, explicable only on the basis of the particular behavior of the australopithecines who had lived in the particular cave.

Dart published voluminously on the new fossils as they appeared, and on his new ideas as they developed. He was criticized thoroughly for his ideas of his 'osteodontokeratic' (tooth-bone-horn) culture, on two bases primarily: 1) australopithecines were apes, and apes didn't make tools (or, even if australopithecines could be called the most primitive hominids, they were so small-brained that they could not have evolved a tool-making culture), and 2) the data could be more logically explained by the bone-collecting and bone-breaking behavior of other animals, particularly of hyenas. The first argument was (and is) illogical; one should allow the data to shape the conclusions, and not demand that the data fit prior assumptions. On the second count, the critics were blissfully ignorant of the behavior of the carnivores to whom they attributed bone-collecting; neither hyenas or any other carnivores drag bones into caves, as Dart and his colleagues carefully pointed out for South Africa and as I have been able to verify for hyenas in Nubian Egypt. So, for the time, Dart's hypothesis was the only one offered which explained the facts, and I must say that for myself at first I found it most compelling, but in general (in spite of Howell's pictures in EARLY MAN of australopithecines sitting in their caves and making tools of bone) I believe that most people were unconvinced that australopithecines had made tools, of teeth, bone, horn, or other materials. (Dart, ever imaginative, also wrote on tools of wood, leather, and cordage, all before we knew much about the abilities of the greater apes to use and fashion tools out of various materials).

Beginning even before the 1940s, but intensifying then and carrying on through Dart's writing for two decades thereafter, another theme was consistently sounded: the australopithecines, at least the gracile ones, were active and successful hunters, and meat was important in their diet. They were thus, of necessity, killers of game, as proved by the thousands of broken bones (with a high concentration of small and medium-sized antelopes) found in the living-caves; additionally, the rare bones of their own kind found in those same living-sites (the caves) were always broken, often with marks of violence upon the bones. By implication, thus, australopithecines killed other australopithecines, dismembered the skeletons, and broke the bones, with cannibalism a presumptive factor. From these conclusions emerged the further concept that killing was a way of life which had had selective value in human evolution. From this idea the next short step in the thought-pattern was that selection for killing set behavioral patterns which were inherited, and thus instinctive, explaining the modern propensities for murder and war. These ideas were to be picked up by Robert Ardrey in the early 1960s and expanded further, but that takes us into an era outside our announced boundaries for this essay.

To return, thus, to the time of our chronological end-point, that time of the death of Broom was a period of continuing mental and physical activity for Dart, in the midst of his development of his ideas on the australopithecines as culture-bearing tool makers. He has not abandoned his faith in the validity of this concept in the slightest, merely because few have agreed with him, and fewer do now since C. K. Brain has shown that the South African caves were probably not walk-in, live-in caves, but more probably were underground caverns with connections to the surface by vertical fissures, down which broken bones from leopards' kills, after being reworked by hyenas and other predators, could fall onto a talus-cone below. Dart, cheery as always, once said to me relative to these new ideas tending to erode the basis for his concept of an osteodontokeratic culture, "The world scoffed in 1925 when I published on the hominid likenesses of the Taung specimen, but now they all say how keen was my vision given so few facts. This will be the pattern again when everyone accepts the truth of the osteodontokeratic culture, as I did on the basis of my own research decades ago." More recently he has written me the same message. Eventually the facts, many of them not yet discovered, will point the way to the acceptance on one conclusion or another, or perhaps one not yet advanced. I myself now think that this second time Dart may not be proved correct.

When Le Gros Clark returned to London early in 1947, fired with the convert's zeal to convince his peers in paleo-anthropology of the truth of the long-derided notions of Dart and Broom that the australopithecines had human-like traits because they really were related to humans, he gave talks and wrote papers, and immediately discovered that he had two severe critics. One was F. Wood Jones, who had been claiming for decades that hominids had had their original ancestry in Eocene tarsioids and therefore no fossils having resemblances to monkeys or apes could be hominids or hominid ancestors. Although Jones was an excellent anatomist, few by 1947 took seriously his concept of a tarsioid ancestry for hominids. The other demurrer, and a more serious one with consequences continuing to the present, was Solly Zuckerman. Born in Cape Town in 1904, Zuckerman grew to adulthood in South Africa, and had his education through the master's level at the University of Cape Town before going to England for his medical studies. A dynamic individual, he remained in England for a varied and brilliant career, leading eventually to a lordship. Early in that career (Zuckerman 1928) he determined to his own satisfaction that Australopithecus africanus was an ape, and he has not yet seen sufficient evidence to change his mind. By 1947, when he was arguing with Le Gros Clark at a meeting of the Anatomical Society of London, he had done considerable behavioral, anatomical and physiological research on primates, and before he was thirty had published two books on these subjects.

Zuckerman, rather reasonably one must admit, demanded from the proponents of hominid similarities for the australopithecines some more decisive evidence than the descriptive and pictorial accounts which had been presented by Dart, Broom, Le Gros Clark and others to that time; indeed, he asked for multiple measurements of all specimens and statistical evaluation of those data. In retrospective answer, after a good deal of stormy and murky waters had passed over the intellectual dam, Le Gros Clark (1967, p. 33) had the following to say of this request: ". . .these differences [i.e., between australopithecines and apes], most of them severally and certainly all of them taken in combination with each other, were so sharp and so obtusive on visual comparison as not to need statistical analysis in order to demonstrate that they were well substantiated. . .I drew attention to some significant hominid (and non-pongid) features of the braincase, the forehead, and cheek region of the skull, the shape of the jaws, the mixture of primitive characters (characters of common inheritance) and hominid characters in the capitate bone of the wrist and the talus bone of the ankle, the inference to be drawn from the lower end of the thighbone indicating an habitual upright posture, and, more especially, the essentially hominid characters of the dentition." To Zuckerman and others of like mind, this type of descriptive approach was all too vague, but if the conclusions were valid, that validity must be determined by multiple measurements treated statistically to determine similarities or dissimilarities with degrees of probability. They set themselves the task, therefore, of making the measurements and performing the statistical manipulations.

Before they began, however, Zuckerman and his colleague E. H. Ashton stated as a first principle that the skulls of australopithecines looked like those of apes (even if not exactly like any one kind of known ape), that the endocranial volumes were within the range of those of living apes, the face was prognathic (as it is in all apes), the cheek-teeth were larger than are those of known hominids (as is true also of apes), and the evidence for bipedalism was scant to absent. Except for the last-named character, all paleoanthropologists of the time would have agreed with the above statements, but some would have added (as Dart and Broom had in 1925, as Sollas did in 1925 and 1926, and as Le Gros Clark had in 1947) that the several obvious similarities of the australopithecines to apes was not the problem, but the important characters were those which, alone in the australopithecines, diverged from the pongid pattern toward that of known hominids. No matter that Zuckerman (1950d, p. 447), wrote of such characters as being "often inconspicuous"; the important point was the presence of several such incipient hominid characters in functional combinations. This latter point of view was one which, in my opinion, Zuckerman and his co-workers failed to grasp, even while they stated that they did. Their approach (again, this is my personal opinion) was extremely static in that they essentially demanded that a fossil, to be considered by them to show any evidence of evolving toward living humans, must have essentially arrived at that latter status before they would regard it as having begun the evolutionary journey. Arguing from this base, any population earlier than, and more primitive than _Pithecanthropus_ (=_Homo erectus_) would undoubtedly not have been accepted as hominid.

One mental block that Zuckerman had throughout the controversy over the phylogenetic position of the australopithecines was his belief that the human line, while related to that of apes at an extremely primitive level, had been separate as a distinct family already in the Oligocene; at least he so believed in 1954, after the controversy had been boiling for seven years (Zuckerman 1954, p. 302). In the words of Gregory (1949) he and his supporters were both 'Pithecophobiacs' and 'Vectorians'; they had an emotional repulsion to the thought of being closely related to living apes and they elected thus to draw a long independent line for hominids. If their premise that the hominid line had been independent since sometime in the Oligocene were true, and many did so believe a quarter of a century ago, and if the australopithecines were apes, then any Plio-Pleistocene hominids would have been expected to have been more similar to living men than the australopithecines were.

This particular intellectual tussle, beginning with Le Gros Clark and Zuckerman in 1947, assumed several lines of argument, most of which occurred simultaneously with considerable repetition in numerous articles in different journals, and various other paleoanthropologists and other specialists got involved along the way. I will try here to sort out the major topics for discussion, and treat them separately, as possible.

Endocranial Volume:

Broom had often, with typical enthusiasm (less carefully controlled, I believe, in his elder years), overestimated the endocranial volume (which he often erroneously called 'brain' or 'brain-cast') when a part of a skull or even a large jaw had been excavated. At least one such estimate for a robust australopithecine was expressed as ". . .perhaps well over 1000 c.c." (Broom and Robinson, 1952, p. 9), published a year after his death, but several similar statements are scattered through his multiple short published notes. If any such figure had been true, the endocranial volume would have overlapped the lower end of the range of pithecanthropines considerably and of living humans slightly. Sometimes these excessive estimates were scaled down in later publications, but not always, and in any case they got into popular writing of the period and were then repeated by others.

Ashton went to South Africa and made his own determinations of cranial capacities on the original specimens; he found that many of the fossils about whose 'brain-size' Broom had guessed could not be utilized for volumentric measurements, not even for estimates. His results (Ashton 1950) were much more conservative and, as time has proved, much more accurate than prior tabulations. (In Table No. 1 I have calculated the means, standard deviations, and ranges from Ashton's data.)

(See Table No. 1)

Although the figure of 650 cc. for the endocast of any australopithecine from South Africa is now known to

Table No. 1

Means, Standard Deviations, and Ranges from Ashton's data

	No.	Mean & s.d.		Range @ 2 s.d.	Range @ 3 s.d.
Australopithecus africanus (type)	1	500			
Plesianthropus transvaalensis	4	506	26.7	452.6 - 559.4	426 - 585
Australopithecus prometheus	1	650			
Paranthropus robustus	1	650			
Pan sp. (chimpanzees)	61	384	37.5	309.4 - 459.4	272 - 497
Gorilla gorilla	211	513	49.2	413.3 - 612	364-661

be too high by more than 100 cc. (Holloway 1974), this primary set of evaluations by Ashton served to eradicate the more extreme of the wild surmises from the scientific literature, but at the same time led to the now-outmoded idea that 800 cc. was a valid 'cerebral rubicon,' a line drawn across time in the evolutionary sequence, with all hominids (including pithecanthropines) having more cranial volume than the magic datum of 800 c.c., but any primate with less cranial capacity being automatically excluded from the Hominidae. This idea came originally from Keith (1948, p. 205) and was then elaborated upon by Vallois (1954). Whereas Keith was using the figure of 800 c.c. as a mean, for a hypothetical group of evolving primates passing from non-human to human status, Vallois interpreted it as the limit of the ranges of the two populations, hominids above and non-hominids below, and presumably meant that limit to be a major criterion at the family level in the classification of any fossil skull.

Zuckerman, in his several reviews (1950d, 1952, 1954) used the figures for cranial capacities in the same spirit as Vallois did; australopithecines, whose range of cranial capacities (435-650, as then believed) fell within the combined range (290-685) of chimpanzees and gorillas, were automatically classified as apes.

No consideration was given to the possibilities that the brains of australopithecines, for all of their small size, had external characteristics which indicated evolution toward living hominids. For instance, Dart's original observation (1925a) that the lunate and parallel sulci were further apart in the type of *Australopithecus africanus* than on any ape's brain, thus denoting superior powers of association, was ignored. Perhaps Schepers' imaginative reconstructions of external surfaces of australopithecine brains from endocasts (Schepers 1946, and repeated 1950) discouraged all similar reports of observations and conclusions, and certainly Zuckerman's participation in the research on endocranial casts of chimpanzees (Le Gros Clark et al. 1936) must have influenced his opinions, but I believe that Ashton and Zuckerman would have classified the australopithecines as apes on the basis of size of endocasts alone.

The various fallacies in using cranial capacities or supposed brain weights alone as evidence for intellectual abilities and thus of evolutionary and/or taxonomic status were exposed some years later (Dart 1956), using multiple examples of different kinds of neurological data, expressed in simple language and without statistics.[36] Except as a historical curiosity, one has not heard of the 'cerebral rubicon' since, and of course with the more recent increase in number of fossils from East Africa, the once-existing gap between cranial capacities of australopithecines and early members of Homo is becoming non-existent.

Dentition:[37]

During 1950, Ashton and Zuckerman (1950a, b) and Zuckerman alone (1950a, b, c) published a series of articles on the dimensions of teeth of the large apes, of the australopithecines, and of the living and fossil hominids (only Homo and the pithecanthropines, as then defined). The three articles by Zuckerman bore the same or similar titles, and were answered by Le Gros Clark (1950a, b, c, d), two of these four articles had the same title as two of Zuckerman's, leading to the possibility of bibliographic confusion. Also that year, Zuckerman (1950d) published a major review article, outlining his interpretation of the data on the dentition as well as all other aspects of australopithecine morphology. His conclusions invariably agreed with those who, like himself, opposed hominid similarities and phylogenetic affinities between australopithecines and 'man.' (By 'man,' the paleo-anthropologists of the day seem usually to have meant post-Neandertal hominids, although this meaning is rarely made clear.)

In this review of the australopithecines, Zuckerman (1950d) did not find one suggested hominid feature which he could accept; even the broad, short ilia, known by that date from Sterkfontein (Broom and Robinson, 1947) and from Makapan (Dart 1949a, b)[38] and undoubtedly the strongest evidence known then or since for bipedal locomotion, were only mentioned but not discussed.

In this preliminary review, Zuckerman (1950d) utilized the data published the same year (Ashton and Zuckerman 1950a, b) to indicate that, to his own satisfaction, the teeth of the different australopithecines could be matched by the teeth of one or another of the Ponginae, and sometimes by more than one, in criteria of gross size, and/or indices between measurements of gross size. These arguments, plus the tables of data, were presented in detail in Ashton and Zuckerman (1950a, b) but even before these publications, Le Gros Clark (1950a) reacted sharply, stating that separate measurements on each tooth, and indices between no more than two or three measurements on each tooth, were not sufficient to demonstrate affinities of populations (the Australopithecinae) which were evolutionarily transitional, and that certainly such individual measurements and indices gave no hint of the total morphological pattern, this latter the result of a long period of selection. Zuckerman responded (1950c) that he was only trying to show that many if not most of the statements about comparative sizes and shapes of teeth in the literature re living great apes, australopithecines, and humans were either erroneous or based upon insufficient data, whereas he and Ashton were building a firm foundation for such studies. Furthermore, Zuckerman wrote, the concept of a 'total morphological pattern' was only an abstraction, a mental construct, that would vary anew in the mind of each new investigator. He then quoted his own measurements, of the kind Le Gros Clark had just stated were invalid, to refute each of the claims of differences of size and shape of the specific teeth which Le Gros Clark (1950a) had cited as showing the uniqueness of australopithecines.

The latter (Le Gros Clark, 1950b) then responded in a way much as, I think, I would have done; since some of the teeth of the australopithecines look totally different than do their homologues in any ape, he showed pictures of the adjacent dc_1 and dp_3 (this latter termed 'first milk molar' in the 1950s)of Homo sapiens and Paranthropus robustus (very similar to each other) and those of 10 young chimpanzees (all similar to each other but totally dissimilar to the same teeth of the first two taxa). He also gave some indices of measurements of canines which did distinguish those teeth of chimpanzees from canines of Paranthropus robustus and Australopithecus africanus. The implication was that measurements alone were not diagnostic; one had to know what to measure. Zuckerman (1950c) answered in turn that he and Ashton had limited themselves to those measurements which Broom and others had already presented and upon which presumably they had based their conclusions; Le Gros Clark's information, thus, was new and so outside the scope of the topic as defined by prior publications. In my opinion, the answer was a weak one, and illustrated what I have mentioned before about the static approach of Zuckerman and his colleagues; faced with obvious visual differences between the teeth of different taxa they chose not to measure those differences.

By late 1951 two pairs of statisticians had taken a look at the data and methodology of Ashton and Zuckerman. Bronowski and Long stated, quite as Le Gros Clark (1950a) had done intuitively, that use of individual measurements and indices was relatively valueless, but instead multivariate analyses must be used. Thus was this technique introduced to paleo-anthropology. As an example, Bronowski and Long (1951, 1953) used multivariate analyses on measurements of lower deciduous canines of modern humans, of australopithecines, and of chimpanzees and showed that these teeth of australopithecines fell completely in the human range and totally outside that of chimpanzees, as Le Gros Clark (1950b) had shown with pictures. Yates and Healy (1951), in their article, stressed that multivariate analyses were fine but not always necessary; indeed, some of the measurements of Ashton and Zuckerman, contrary to their published statements, and on the bases only of means and standard variations, did distinguish between some teeth of apes and those teeth of australopithecines. This discrepancy between fact and published error was due to a consistent mistake by Ashton and Zuckerman in their calculations of the standard deviations of the individual measurements on all teeth, a mistake which the latter promptly acknowledged (Ashton and Zuckerman 1951b, 1952 a,b).

Le Gros Clark had also been busy measuring teeth,[39] and published the results of his study in 1952.[40] His statistics amounted to no more than ranges, indices, means, and standard deviations for measurements on jaws and several of the teeth of modern humans, chimpanzees, gorillas, and specimens from four of the five known australopithecine sites (Makapan was not included). Usually, where morphological differences were so obvious as he thought to be apparent to all, only drawings were included (i.e., shapes of dental arcades as between apes and modern men). Where the measurements taken by Ashton and Zuckerman did not reveal, but instead hid, differences obvious to the eye, Le Gros Clark in some instances devised measurements and indices which yielded statistical confirmation of the visually obvious. Upon consideration of his results, Le Gros Clark found overwhelming evidence to state that australopithecine dentitions were much more hominid-like than they were ape-like, the truth of which he had been convinced before he began the detailed measurements. In an Addendum, Le Gros Clark called attention to the statistical inadequacies of the miscalculations of the standard deviations as published by Ashton and Zuckerman, and stated, "...the standard deviations given by Ashton and Zuckerman in their study of the teeth of anthropoid apes were incorrectly calculated, and, as the result of this error, the conclusions based on their comparative studies are no longer valid."

Ashton and Zuckerman did not even acknowledge the possibility of the above conclusion by Le Gros Clark; recalculation of their standard deviations on the individual teeth (Ashton and Zuckerman 1952a,b) did narrow the range of this statistic, thus in some cases negating their prior claims that any tooth of any australopithecine could be matched in any measurement of index with a tooth of a greater ape. This shift of evidence did not change Zuckerman's conclusions, as in his next major review (1952) on the status of australopithecines, he wrote, "...few of the australopithecine teeth that have been compared differ significantly from corresponding teeth of the orang."

I have here reviewed in some detail the opposing views of Le Gros Clark and Zuckerman, particularly, to show that, given the same specimens, different people can take some similar and some different measurements and come to completely opposing conclusions, with and/or without the aid of statistics. Each antagonist emerged from the fray 'proving' the conclusion he had had in advance; my own history had been one of acceptance of the hominid affinities of the australopithecines as early as 1940, so my sympathy in reviewing this literature has been with Le Gros Clark, and, try as I may, I cannot find a sense of validity to most of the assertions and conclusions of Ashton and Zuckerman. The opinion of the majority of paleo-anthropologists, beginning in this period of the late 1940s and early 1950s which we have been discussing, has swung over to agree with Le Gros Clark, and thus with Dart, Broom, Sollas, Eliot Smith, Adloff, Gregory and Hellman, Senyuruk, Robinson, Tobias, Washburn, Howells, and an increasing number of specialists who no longer even consider the possibility that australopithecines are apes.

With this latter decison, Zuckerman continued to disagree, and, insofar as I can determine, continues to disagree. Even before he had finished the series of articles on endocasts and dentitions of hominoids, as listed in the bibliography to this chapter, he had launched a series of investigations of posture and locomotion as correlated with the osteology of the pelvic limbs, of the shape of the head and the growth of the skull, of the relative position of the occipital condyles, of the sagittal crests, of the mastoid processes, of the basicranial axes, of the mandibles, and of the mandibular regions. Much of this material was reviewed in 1954, but some articles continued to appear for the next few years. We are beyond the time-limit set for this chapter; Broom had been dead since early in 1951 and Dart in the mid-fifties was busy with the details of his study of the osteodontokeratic culture. Still Zuckerman and his colleagues, no matter what the measurements or how those data were analyzed, for the most part continued to find again the same conclusion that the australopithecines were more similar to apes than to hominids.

One might think, with all of the data pointing to the same conclusion, year after year and character after character, that Zuckerman et al. must inevitably have been right. If they were then, they are now, but few would agree, and the accumulating evidence of multiple new-found fossils continue to remove any possibility for such agreement.

There is a lesson in all of this, which I will allow each reader to interpret.

NOTES

[1]For a bibliography of Raymond Arthur Dart through 1968 see Fischer (1969).

[2]Taung is in that part of the Cape Province of South Africa formerly, and sometimes still, called 'British Bechuanaland' (or simply 'Bechuanaland'); it is not in the former 'Bechuanaland Protectorate,' now the independent nation of Botswana. Confusion between the two 'Bechuanalands' has sometimes led to the erroneous placing of Taung in the Protectorate.

[3]Tobias (1968) has stated that the fossil or the endocast alone (the wording is not clear) had served for an unknown period of time as a paper-weight on the desk of the manager of the Buxton Quarry. I suggest that while serving this utilitarian function the two parts, face and endocast, were still attached; otherwise the non-functioning piece might well have been lost or discarded. In his original publication on _Australopithecus_, Dart (1925a) mentioned the "... fractured frontal extremity..." of the endocast, by which it articulated exactly with the facial portion. Thus the two parts, separate when Dart first saw them, may have been broken apart only when packing them for shipment or in transit.

[4]I have heard that in the 1870's and/or 1880's, Edward Drinker Cope, excavating by day and writing by lantern at night in the fossil fields of the North American 'High Plains,' 30 miles from the nearest telegraph station, could have his article or articles published the next day in Philadelphia. In the mid-1920's, Dart's article, sent from South Africa to London by sea-mail, was published in less than seven weeks after he had finished exposing the face of the fossil. In today's period of instant world-wide communication, one is fortunate to get an article published in a year or even two. Ah, progress!

[5]Dart's claim for bipedalism depended upon the anterior position of the foramen magnum. At the time of its discovery, Dart's type specimen had some of the right exoccipital still in place beside the base of the cast of the cerebellum. From the position of this piece of exoccipital, Dart determined the position of basion. Since mention of this piece of occipital appears only rarely in subsequent writings on the subject, I wrote Dart and asked him

about it. He answered that his original description was correct. To clear any misconceptions, I quote here that part of his letter related to the present condition of this piece of occipital:"...after information by Mr. Alun Hughes and subsequent personal inspection of the skull at our Medical School (where it is now housed under lock and key) ..." Dart determined that "...about a centimetre length of the condylar part of the occipital bone is present still as far back as the jugular notch curvature and transverse sinus of the brain case. It's the basilar portion (that does not fuse until the 18-20 year period in man) that has vanished, though I recall a piece of it also.

Please remember too that the specimen has gone through the hands of numerous inquisitive investigators, sight-seers and cast makers since I described what was there over half a century ago. So the answers are: (a) a part of the right exoccipital was and still is present; (b) having reconstructed the brain and the skull with its occipital opening from casts and the data available, I took my basion measurements from it."

[6]While casts were not available in February, 1925, Dart did have casts prepared as rapidly as possible for the British Empire Exhibition of 1925, held that summer in England (Dart 1967, p. 41). At least one cast had reached London by June 9, 1925, for on the evening of that date Professor G. Elliot Smith exhibited, and made remarks upon, a cast of *Australopithecus africanus* (Proceedings of the Zoological Society of London, 1925, p. 1238). Sometime prior to June 22, 1925, the several casts which Dart had sent to England were on display at Wembley (Keith 1925c), and Keith complained that the casts were not available for purchase; indeed, at the time they were not available for study or even close inspection, as they were being exhibited behind glass. Presumably, thus, only one set of casts had been sent to England at the time.

An additional factor in obscuring the details of the endocast when found, and for some two decades thereafter, was that a considerable part was covered by a thin layer of bone, the inner layer (plate) of the adjacent roofing bones. This fossilized bone was, with Dart's permission, removed by Dr. G.W.E. Schepers (1946, pp. 167-168) prior to the latter's own detailed study of the endocast. Casts made before the removal of this thin shell of bone and those made afterwards are obviously different in detail.

[7]In addition to having received honors in Geology as an undergraduate, Broom had made many geological observations in Australia and South Africa. While the number of his purely geological publications by 1925 was only four, his numerous paleontological articles often contained geological notes. Additionally, for seven years (1903-1909) Broom was Professor of Zoology and Geology at Victoria College, University of Stellenbosch. He resigned his position there amidst praise from all as to his scholarship and ability as a teacher, but the pay was low and the teaching interfered with his research on mammal-like reptiles. My own opinion is that Broom may have been increasingly disturbed by the routine.

[8]The phrase 'human occupation' in the title of Dart 1925d was not based on recovery of artifacts or hominid bones at Makapan at that time, but instead upon the fact that some of the bones in the breccia had the appearance of being charred. Chemical testing indicated the presence of free carbon; on the basis of the carbon the presence of fire-tending hominids was assumed. When bones of australopithecines, almost a quarter of a century later, were recovered from the same breccias, the specific name prometheus (Dart 1948) was bestowed upon the population, not only because of the earlier mental association of the site with fire but additionally a vitreous material, which must have been formed in fire, was found beneath the deepest deposits in the cave. Oakley (1954) has conjectured, however, that the vitreous material was formed when considerable accumulations of guano from bats caught fire and burned under natural conditions.

[9]One interesting item in Dart's article of 1926, not noted otherwise that I remember, is the following: "...fragments of the distal ends of the arm bones and of the phalanges were present in the rock mass from which the facial fragment was isolated."

[10]At the annual meeting of the American Association of Physical Anthropologists held in Boston, April, 1971, a memorial was read in memory of William King Gregory, who had died the previous December at the age of 96. Two graduate students sitting behind me were heard asking each other, "Who was William

King Gregory?" Gregory, who was as much at home with fossil fishes as with living fishes and with fossil Primates as with living Primates (and with all other vertebrates, too), was twice president (1941-43) of the American Association of Physical Anthropologists and in 1949 was also the recipient of the Viking Fund Medal in Physical Anthropology. He deserves better from a new generation of graduate students than total oblivion.

[11]In contrast to this indication by Gregory of the importance of *Australopithecus* in hominid evolution, his equally famous colleague at the American Museum of Natural History, Henry Fairfield Osborn, writing on the same subject (hominid evolution) in the same year and the same journal, did not even mention *Australopithecus*.

[12]The casts were prepared and sold by Damon and Co., London, as were many of the casts of fossils, hominid and others, commercially available at the time. A brief history of this company, and of the eventual disposal of their casts and molds to the University of Pennsylvania, has been recounted by Keith (1952). I have not been able to learn if the casts Abel used had been made from the original specimen of *Australopithecus africanus* after Mrs. Dart arrived with it in London in 1930, or whether they were copies of casts Dart had earlier distributed (1925 and 1929) from Johannesburg.

[13]One can only suspect some degree of paternal influence, even if unconscious.

[14]Gregory's article of June, 1930, probably reached Vienna too late to be seen by Abel before his own manuscript had been sent to the printer.

[15]Broom had given the Croonian lecture in London at the Royal Society in 1913; he was elected a member of that Society in 1920; he had received the Royal Medal of the same Society in 1928; his book "Mammal-like Reptiles of Southern Africa" had received world-wide acclaim; and he had, as mentioned, been president of the South African Association for the Advancement of Science in 1933.

[16]For comparison, consider that the same year (1934-1935), I attended the University of Oregon as an undergraduate for $300 for all expenses. Broom's

salary for the year came to approximately $2,500, as much as many associate professors in the United States were earning then.

[17]The casual taxonomic practice of the day is well-illustrated by the introduction of transvaalensis to the scientific world as the name of a new species. In the description of the specimen Broom mentioned that he regarded it as representing a new species, but no name was proposed in the text. Instead, the name -- not designated as new -- is to be found only in the caption to fig. 4.

[18]The practice in paleontology, and particularly in the paleontology of Primates, of increasing the number of names of genera and species beyond the limits of all biological probability was not due entirely to ignorance. Franz Weidenreich (1946, p.3) stated the issue clearly, in answer to critics who had recommended that he and other paleo-anthropologists change their Linnean nomenclature to fit the facts of biology: "...I have refused. In paleontology it always was and still is the custom to give generic or specific names to each new type without much concern for the kind of relationship to other types formerly known. Furthermore, the old names of fossil human types are accepted throughout the entire literature dealing with early man, so that any radical change would lead to the greatest confusion and necessitate complicated explanations in each case. I shall continue, therefore, to use the old names without imputing a special taxonomic meaning." I suspect that Weidenreich's defense of the continuing errors of the past re misue of systematic principles in primate paleontology was induced by G.G. Simpson's stinging criticism of that misuse the previous year (Simpson 1945, p. 188).

[19]G.G. Simpson (1945, p. 187) has written of apes and men together (his Superfamily Hominoidea), "Wilder boldy united all these forms in a single family, Hominidae, and Gregory and Hellman have adopted this arrangement." Simpson quoted Gregory and Hellman 1939b, as designated in this chapter, as their publication in which they united the apes and men into the Hominidae. I do not find reason for this conclusion in my reading of the same publication.

[20]About 1940, then a graduate student, I began to keep a cross-referenced bibliographic file of articles I was reading. Without any profound thoughts on the subject, and without having heard of Adloff, I included the Australopithecinae in the Hominidae. I did not then know how alone I was in that practice, but I was influenced by hearing about the research of Gregory and Hellman, without at that time having read any of their articles.

[21]The terminology of the sites in the Makapan Valley is confusing. Makapan was a Bantu chief of the first half of the Nineteenth Century; 'gat' in Afrikaans means 'hole' or 'cave.' 'Makapansgat' is a cave in which Makapan and his Bantu followers underwent a siege by Boers in 1854. That cave has since been declared a historic monument and is now officially named the 'Historic Cave,' although still sometimes still called 'Makanpansgat.' Two other caves, close to the Historic Cave and both containing Paleolithic artificts, were subsequently opened by lime-miners and named 'Cave of Hearths' and 'Rainbow Cave' (van Riet Lowe 1948, fig. 1). No fossils of australopithecines have been found in any of these three caves; all such fossils in the Makapan Valley have come from the cave-fill of a commercial lime-works hardly more than a half-mile west, as the hoopoe flies, but almost a mile by road. This fossil-bearing site, generally called the 'Limeworks Site' originally, is now usually termed 'Limeworks' or 'Makapan,' but should not thereby be confused with the original Makanpansgat (= Historic Cave). (See the map in Dart 1967, p. 90, and also see Barbour 1949a,b).

[22]The observations by Le Gros Clark et al. (1936) on the relations, if any, between the positions of sulci and gyri on a brain and corresponding grooves and ridges on the endocast of the same individual chimpanzee has indirect pertinance, not only to Schepers' observations and deductions, but to those made originally by Dart on the endocast of the type specimen of Australopithecus africanus. Obviously, an australopithecine is not a chimpanzee, yet the endocranial capacities of individuals of the two populations are quite similar; if anything, the australopithecine often being the larger, should probably have an endocast less distinct than is that of a chimpanzee. The original of Le Gros Clark et al. (1936) should be read in detail by anyone interested, but in general I can say that, with a sample of six

specimens, those authors often (but not always) had difficulty in finding the lunate sulcus on the endocast, but instead found that an endocranial furrow related to the lambdoid suture could be readily mistaken for the lunate sulcus, in which case the inference would follow that the lunate sulcus had acquired the disposition characteristic of some human brains, in that the parietal lobe would be thought to be much larger than in reality it was. Indeed, Le Gros Clark et al. (1936) in their concluding remarks strongly suggested that Dart's interpretation was incorrect and that what he declared to be the 'lunate sulcus' was instead the mark of the lambdoid suture, and that the true lunate sulcus may have occupied a more anterior position, very much as in a chimpanzee.

[23]The name of Wilfred E. Le Gros Clark is often catalogued in bibliographies as 'Clark, W.E. Le Gros,' even by his British contemporaries, but in his own later books which I have he has listed himself under 'L' as Le Gros Clark, W.E., and I choose to do likewise.

[24]Although Leakey was almost a half-century younger than Broom, the two men admired each other greatly. Broom particularly liked Leakey's tendency to upset the established order with new discoveries and new interpretations.

[25]The Navy's major interest, let it be said to their credit, was in military medicine, and much of the emphasis of the group was on the practical aspects of the study of tropical medicine under field-conditions (Phillips 1949).

[26]Phillips attended the Pan-African Congress in Nairobi in January, 1947, and published a report on it (Phillips 1947). He had the ability to write a sensible, straight-forward scientific report when he wished to do so, but as he got older and richer his level of writing became shallower and more romanticized.

[27]In the short note on the blue buck Broom promised that a full account of the skull would be published later. Presumably he had himself in mind as the author, but this time Broom erred. The description *was* published later, by 18 years, but the author was the German mammalogist, Erna Mohr (1967).

[28]None of these three authors mentioned Adloff, who had already made the same assignment of *Australopithecus* to the Hominidae in 1932.

[29]Mayr has many capabilities, but he is not a paleontologist, as Dart once identified him (Dart 1967, p. 207).

[30]In a talk I heard Mayr give on this topic in the spring of 1950, he said simply that all primates differed no more among themselves than did all species of *Drosophila* among themselves. By the time he arrived at the Cold Spring Harbor Symposium that summer he must have received enough adverse criticism to have modified his statement to the extent of suggesting the inclusion of only the Suborder Anthropoidea in *Homo*. I have, myself, never been certain how serious Mayr was with his suggestion, but I have always suspected that he really wanted to make the point that hominids at the time were cursed with a vast over-supply of generic and specific names.

[31]Mayr gave no indication of a reason for using '*transvaalensis*' as a specific name for all australopithecines, when '*africanus*' clearly had and has priority. He was too good a systematist to have considered discarding the name '*africanus*' because it was based upon an immature specimen with which other fossils could not easily be compared -- a fallacy temporarily adopted by Tobias several years later (Tobias 1973b, 1974; Olson 1974). Mayr may have been aware that '*africanus*' had, earlier than 1925, been used in a taxonomic sense for a population of living *Homo*, thus invalidating its later use for a different population of *Homo*. If so, Robinson (1972, pp. 8-11) has since demonstrated that this earlier use of '*africanus*' was not itself systematically valid.

[32]Broom had earlier (1940) suggested the possibility of the pithecanthropines being called *Homo erectus*, but only for the purpose of arguing against the idea. Both Mayr and Evans, however, were sincere and their suggestion has generally now been adopted, but paleo-anthropologists in some few countries, for reasons of historic pride, still insist on retaining *Pithecanthropus* as a genus.

[33]A present tendancy exists to list the robust australopithecines from East Africa as Australopithe*cus boisei*, for no reason that I can find. If that population of robust australopithecines is different

enough from those in South Africa to warrant being awarded separate specific status, the present native human population of sub-Saharan Africa would have to be divided into at least four species.

[34]Broom had urged Dart to give a new generic name to the new population of australopithecines the latter was describing from Makapan, and was sincerely disappointed when Dart continued with the old name *Australopithecus*.

[35]I find the first expression of this opinion that the bone-breccias of australopithecine cavesites resulted from occupation by living australopithecines who purposely accumulated and there fragment the bones of numerous animals in Dart (1926), when only the single site of Taung was known. He expanded upon this theme in 1929, and returned to it many times thereafter; indeed, at the time of present writing (1978) he has not changed his mind about the nature of the origin of the bone-breccias of the South African 'caves,' in spite of much opposing evidence.

[36]This article should be read by all because of its fascination.

[37]The numerous publications on the comparisons of australopithecine teeth with those of other hominoids should be read in chronological sequence to get the full flavor of the developing arguments. Insofar as I can determine, that sequence is: Le Gros Clark 1950a; Zuckerman 1950a; Le Gros Clark 1950b; Zuckerman 1950b; Ashton and Zuckerman 1950a,b; Le Gros Clark 1950c,d; Zuckerman 1950c,d; Le Gros Clark 1951a; Zuckerman 1951a; Le Gros Clark 1951b; Zuckerman 1951b; Ashton and Zuckerman 1951a; Keith 1951; Bronowski and Long 1951; Yates and Healy 1951; Ashton and Zuckerman 1951a; Le Gros Clark 1952; Ashton and Zuckerman 1952a,b; Zuckerman 1952; Browowski and Long 1952. Teeth were also discussed by different authors in later papers, but those were usually reviews on many topics, as Zuckerman 1954.

[38]A partial innominate of a robust australopithecine from Swartkrans was discovered that same year (Broom and Robinson 1949b), but was published too late to have been noted by any of the participants in the present discussion.

[39]Presumably, the teeth that Le Gros Clark measured were in large part the same teeth that Ashton

and Zuckerman had already measured; the numbers of teeth of apes and of casts of australopithecines available in England were limited.

[40]This publication was dated 1950, and thus is usually listed as of that year in bibliographies, but in the 'addendum' Le Gros Clark referred to publications of December, 1951, and in the bibliography of his book of 1967 he listed the date of publication of the article as 1952.

References

Abel, Wolfgang. 1931. Kritische Untersuchungen über Australopithecus africanus Dart. Gegenbauer's morphologische Jahrbuch 65:539-640.

Adloff, P. 1932. Das Begiss von Australopithecus africanus Dart. Einige ergänzende Bemerkungen zum Eckzahnproblem. Zeitschrift fur Anatomie und Entwicklunggeschichte 97:145-156.

Anonymous. 1947. The Pan-African Congress on Prehistory. Nature 159:216-218.

Anonymous. 1966. The great Iam. Time 88 (30 Sept.): 110-112.

Ashton, E.H. 1950. The endocranial capacities of the Australopithecinae. Proceedings of the Zoological Society of London 120:715-720.

Ashton, E.H., and S. Zuckerman. 1950a. Some quantitative dental characteristics of the chimpanzee, gorilla and orang-outang. Philosophical Transactions of the Royal Society of London 234B:471-484.

Ashton, E.H., and S. Zuckerman. 1950b. Some quantitative dental characters of fossil anthropoids. Philosophical Transactions of the Royal Society of London 234B:485-520.

Ashton, E.H., and S. Zuckerman. 1951a. Some dimensions of the milk teeth of man and the living great apes. Man 51:23-26.

Ashton, E.H., and S. Zuckerman. 1951b. Statistical methods in anthropology. Nature 168: 1117-1118.

Ashton, E.H. and S. Zuckerman. 1952a. Overall dental dimensions of hominoids. Nature 169:571-572.

Ashton, E.H. and S. Zuckerman. 1952b. The measurement of anthropoid teeth, with particular reference to the deciduous dentition. Man 52:64.

Barbour, George B. 1949a. Makapansgat. Scientific Monthly 69, no. 3;141-147.

Barbour, George B. 1949b. Ape or man? An incomplete chapter of human ancestry from South Africa. Ohio Journal of Science 49:129-145.

Bronowski, J., and W. M. Long. 1951. Statistical methods in anthropology. Nature 168:749.

Bronowski, J., and W. M. Long. 1952. Statistics of discrimination in anthropology. American Journal of Physical Anthropology, n.s. 10:385-394.

Bronowski, J., and W. M. Long. 1953. The Australopithecine milk canines. Nature 172:251.

Broom, R. 1925a. Some notes on the Taungs skull. Nature 115:569-571.

Broom, R. 1925b. On the newly discovered South African man-ape. Natural History 25:409-418.

Broom, R. 1929. Note on the milk dentition of Australopithecus. Proceedings of the Zoological Society of London 1929:85-88.

Broom, R. 1930. The age of Australopithecus. Nature 125:814.

Broom, R. 1936a. A new fossil anthropoid skull from South Africa. Nature 138:486-488.

Broom, R. 1936b. The dentition of Austalopithecus. Nature 138:719.

Broom, R. 1937a. The Sterkfontein ape. Nature 139: 326.

Broom, R. 1937b. Discovery of a lower molar of Australopithecus. Nature 140:681-682.

Broom, R. 1937c. On Australopithecus and its affinities. In Early Man as Depicted by Leading Authorities. George Grant MacCurdy, ed., pp. 285-292. Philadelphia: J. B. Lippincott Company.

Broom, R. 1938a. The Pleistocene anthropoid apes of South Africa. Nature 142:377-379.

Broom, R. 1938b. More discoveries of Australopithecus. Nature 141:828-829.

Broom, R. 1939a. The dentition of the Transvaal Pleistocene Anthropoids, *Plesianthropus* and *Paranthropus*. *Annals of the Transvaal Museum* 19:303-314.

Broom, R. 1939b. On the affinities of the South African Pleistocene anthropoids. *South African Journal of Science* 36:408-411.

Broom, R. 1940. Classification of sub-human types. *Nature* 146:94.

Broom, 1943. An ankle-bone of the ape-man *Paranthropus robustus*. *Nature* 152:689-690.

Broom, R. 1946. The South African fossil ape-men: The Australopithecinae. Part 1. The occurrence and general structure of the South African ape-men. *Transvaal Museum Memoir* 2:7-153.

Broom, R. 1947. Discovery of a new skull of the South African ape-man, *Plesianthropus*. *Nature* 159:672.

Broom, R. 1949a. Another new type of fossil ape-man. Nature 163:57.

Broom, R. 1949b. The extinct blue buck of South Africa. *Nature* 164:1097-1098.

Broom, R. 1950. *Finding the Missing Link*. London: Watts and Co.

Broom, R., and J.T. Robinson. 1947. Further remains of the Sterkfontein ape-man, *Plesianthropus*. *Nature* 160:430-431.

Broom, R. and J.T. Robinson. 1949. A new type of fossil man. *Nature* 164:322-323.

Broom, R. and J.T. Robinson. 1950a. Sterkfontein apeman *Plesianthropus*. Part 1. Further evidence of the structure of the Sterkfontein ape-man *Plesianthropus*. *Transvaal Museum Memoir* 4:11-84.

Broom, R. and J.T. Robinson. 1950b. Notes On the pelves of the fossil ape-men. *American Journal of Physical Anthropology* n.s. 8:489-494.

Broom, R. and J.T. Robinson. 1952. Swartkrans ape-man: *Paranthropus crassidens*. *Transvaal Museum Memoir* 6:1-123.

Camp, Charles L. 1948. University of California African expedition - southern section. Science 108:550-552.

Cooke, H.B.S. 1960. A remarkable forecast from the Taungs skull by Dr. Broom in 1925. South African Journal of Science 56:75-77.

Dart, Raymond A. 1925a. Australopithecus africanus: The man-ape of South Africa. Nature 115:195-199.

Dart, Raymond A. 1925b. The word "Australopithecus" and others. Nature 115:875.

Dart, Raymond A. 1925c. The Taungs skull. Nature 116:462.

Dart, Raymond A. 1925d. A note on Makapansgat: A site of early human occupation. South African Journal of Science 22:454.

Dart, Raymond A. 1926. Taungs and its significance. Natural History 26:315-327.

Dart, Raymond A. 1929a. A note on the Taungs skull. South African Journal of Science 26:648-658.

Dart, Raymond A. 1929b. Anthropology. In South Africa and Science: A Handbook under the Auspices of the South African Association for the Advancement of Science. H. J. Crocker and J. McCrae, eds., pp. 267-286. Johannesburgh: Hortors Limited.

Dart, Raymond A. 1934. The dentition of Australopithecus africanus. Folia Anatomica Japonica 12:207-221.

Dart, Raymond A. 1948. The Makapansgat proto-human Australopithecus prometheus. American Journal of Physical Anthropology n.s. 6:259-283.

Dart, Raymond A. 1949a. The first pelvic bones of Australopithecus prometheus: Preliminary note. American Journal of Physical Anthropology n.s. 7:255-257.

Dart, Raymond A. 1949b. Innominate fragments of Australopithecus prometheus. American Journal of Physical Anthropology n.s. 7:301-333.

Dart, Raymond A. 1956. The relationships of brain size and brain pattern to human status. South African Journal of Medical Science 21:23-45.

Dart, Raymond A. 1957. The osteodontokeratic culture of Australopithecus prometheus. Transvaal Museum Memoir 10:1-105.

Dart, Raymond A. 1967. Adventures with the Missing Link. Philadelphia: The Institutes Press.

Duckworth, W.L.H. 1925. The fossil anthropoid ape from Taungs. Nature 115:236.

Du Toit, Alex. L., ed. 1948. Robert Broom Commemorative Volume. Cape Town: The Royal Society of South Africa.

Evans, F. Gaynor. 1945. The names of fossil men. Science 102:16-17.

Findley, George. 1972. Dr. Robert Broom, F.R.S. Palaeontologist and Physician, 1866-1951. Cape Town: A.A. Balkema.

Fischer, Ilse. 1969. Professor Raymond Arthur Dart: A Bibliography of his Works. Johannesburg: Department of Bibliography, Librarianship and Typography; University of the Witwatersrand.

Genet-Varcin, E. 1969. À la Recherche du Primate: Ancêtre de l'Homme. Paris: Boubee.

Gregory, William K. 1930. The origin of man from a brachiating anthropoid stock. Science 71:645-650.

Gregory, William K. 1949. The bearing of the Australopithecinae upon the problem of man's place in nature. American Journal of Physical Anthropology n.s. 7:485-512.

Gregory, William K., and Milo Hellman. 1938. Evidence of the Australopithecine man-apes on the origin of man. Science 88:615-616.

Gregory, William K. and Milo Hellman. 1939a. Fossil man-apes of South Africa. Nature 143:25-26.

Gregory, William K., and Milo Hellman. 1939b. The dentition of the extinct South African man-ape Australopithecus (Plesianthropus) transvaalensis Broom. A comparative and phylogenetic study. Annals of the Transvaal Museum 19:339-373.

Gregory, William K., and Milo Hellman. 1939c. The South African fossil man-apes and the origin of the human dentition. Journal of the American Dental Association 26:558-564.

Gregory, William K., and Milo Hellman. 1940. The upper dental arch of Plesianthropus transvaalensis Broom, and its relations to other parts of the skull. American Journal of Physical Anthropology, n.s. 26:211-228.

Holloway, Ralph L. 1974. The casts of fossil hominid brains. Scientific American 231, no. 1:106-115.

Howell, F. Clark. 1965. Early Man. New York: Life Nature Library; Time Incorporated.

Howell, F. Clark. 1973. Early Man, Second Edition. New York: Life Nature Library; Time Incorporated.

Howells, W.W. 1950. Origin of the human stock: Concluding remarks of the chairman. Cold Spring Harbor Symposia on Quantitative Biology 15:79-86.

Hrdlička, Aleš. 1925. The Taungs ape. American Journal of Physical Anthropology 8:379-392.

Hrdlička, Aleš. 1930. The skeletal remains of early man. Smithsonian Miscellaneous Collection 83: 1-379.

Hrdlička, Aleš. 1935. Yale fossils of anthropoid apes. American Journal of Science, ser. 5, vol. 29:34-40.

Keith, Arthur. 1925a. The fossil anthropoid ape from Taungs. Nature 115:234-235.

Keith, Arthur 1925b. The new missing link. British Medical Journal 1925, pt. 1:325-336.

Keith, Arthur.1925c. The Taungs skull. Nature 116:11.

Keith, Arthur. 1925d. The Taungs skull.Nature 116:462-463.

Keith, Arthur. 1947. Australopithecinae or Dartians? Nature 159:377.

Keith, Arthur. 1948. A New Theory of Human Evolution. London: Watts.

Keith, Arthur. 1951. The dentition of the Australopithecinae. Man 51:70.

Keith, Arthur. 1952. Frank Oswell Barlow: 4 October 1880 - 12 November 1951. Man 52:70.

Krogman, Wilton Marion. 1976. Fifty years of physical anthropology: The men, the material, the concepts, the methods. Annual Review of Anthropology 5:5-14.

Kurtén, Björn. 1972. Not from the Apes. New York: Vintage Books of Random House.

Le Gros Clark, W. E. 1934. Early Forerunners of Man: A Morphological Study of the Evolutionary Origin of the Primates. Baltimore: William Wood and Company.

Le Gros Clark, W. E. 1938. The endocranial cast of the Swanscombe bones. Journal of the Royal Anthropological Institute of Great Britain and Ireland 68:61-67.

Le Gros Clark, W. E. 1946. Immediate problems of human palaeontology. Man 46:80-84.

Le Gros Clark, W. E. 1947a. Observations on the anatomy of the fossil Australopithecinae. Journal of Anatomy 81:300-332.

Le Gros Clark, W. E. 1947b. The importance of the fossil Australopithecinae in the study of human evolution. Science Progess 35:377-395.

Le Gros Clark, W. E. 1950a. New paleontological evidence bearing on the evolution of the Hominoidea. Quarterly Journal of the Geological Society of London 105:224-264.

Le Gros Clark, W. E. 1950b. South African fossil hominoids. Nature 165:893-894.

Le Gros Clark, W. E. 1950c. New discoveries of the Australopithecinae. _Nature_ 166:758-760.

Le Gros Clark, W. E. 1950d. South African fossil hominoids. _Nature_ 166:791-792.

Le Gros Clark, W. E. 1951a. Comments on the dentition of the fossil australopithecines. _Man_ 51:18-20.

Le Gros Clark, W. E. 1951b. The dentition of the fossil australopithecines. _Man_ 51:32.

Le Gros Clark, W. E. 1952. Hominid characters of the australopithecine dentition. _Journal of the Royal Anthropological Institute of Great Britain and Ireland_ 80:37-54.

Le Gros Clark, W. E. 1967. _Man-apes or Ape-men: The Story of Discoveries in Africa_. New York: Holt, Rinehart and Winston, Inc.

Le Gros Clark, W. E., D. M. Cooper, and S. Zuckerman. 1936. The endocranial cast of the chimpanzee. _Journal of the Royal Anthropological Institute of Great Britain and Ireland_ 66:249-268.

Lucas, F. A. 1925. The word "_Australopithecus_" and others. _Nature_ 116:315.

Mayr, Ernst. 1944. On the concepts and terminology of vertical subspecies and species. _National Research Council Bulletin_ 2:11-16.

Mayr, Ernst. 1950. Taxonomic categories in fossil hominids. _Cold Spring Harbor Symposia on Quantitative Biology_ 15:109-118.

Oakley, Kenneth P. 1954. Evidence of fire in South African cave deposits. _Nature_ 174:261-262.

Olson, T. R. 1974. Taxonomy of the Taung skull. _Nature_ 252:85.

Osborn, Henry Fairfield. 1930. The discovery of Tertiary man. Science 71:1-7.

Peabody, Frank E. 1954. Travertines and cave deposits of the Kaap escarpment of South Africa, and the type locality of _Australopithecus africanus_ Dart. _Bulletin of the Geological Society of America_ 65: 671-706.

Phillips, Wendell. 1947. The first Pan-African Congress on Prehistory. <u>Science</u> 195:611-613.

Phillips, Wendell. 1949. Africa from Nubia to Turkana. <u>Scientific Monthly</u> 69:262-269.

Remane, Adolf. 1951. Die Zähne des <u>Meganthropus africanus</u>. <u>Zeitschrift für Morphologie und Anthropologie</u> 42:311-329.

Remane, Adolf. 1954. Structure and relationships of <u>Meganthropus africanus</u>. <u>American Journal of Physical Anthropology</u> n.s. 12:123-126.

Robinson, Arthur. 1925. The Taungs skull. <u>British Medical Journal</u> 1925, pt. 1:622.

Robinson, J. T. 1953. <u>Meganthropus</u>, australopithecines and hominids. <u>American Journal of Physical Anthropology</u> n.s. 11:1-38.

Robinson, J. T. 1967. Variation and the taxonomy of the early hominids. <u>Evolutionary Biology</u> 1:69-100.

Robinson, J. T. 1972. <u>Early Hominid Posture and Locomotion</u>. Chicago: University of Chicago Press.

Romer, Alfred S. 1930. <u>Australopithecus</u> not a chimpanzee. <u>Science</u> 71:482-483.

Romer, Alfred S. 1945. <u>Vertebrate Paleontology</u>. Second Edition. Chicago: University of Chicago Press.

Schepers, G.W.H. 1946. The South African fossil ape-men, the Australopithecinae. Part II. The endocranial casts of the South African ape-men. <u>Transvaal Museum Memoir</u> 2:155-272.

Schepers, G.W.H. 1950. Sterkfontein ape-man <u>Plesianthropus</u>. Part II. The brain casts of recently discovered <u>Plesianthropus</u> skulls. <u>Transvaal Museum Memoir</u> 4:85-117.

Senyürük, Muzaffer Suleyman. 1941. The dentition of <u>Plesianthropus</u> and <u>Paranthropus</u>. <u>Annals of the Transvaal Museum</u> 20:293-302.

Simpson, George Gaylord. 1945. The principles of classification and a classification of mammals. Bulletin of the American Museum of Natural History 85:1-350.

Smith, G. Elliot. 1925. The fossil anthropoid ape from Taungs. Nature 115:235.

Sollas, W. J. 1925. The Taungs skull. Nature 115: 908-909.

Sollas, W. J. 1926. On a sagittal section of the skull of Australopithecus africanus. Quarterly Journal of the Geological Society of London 82: 1-11.

Symington, J. 1916. Endocranial casts and brain form: A criticism of some recent speculations. Journal of Anatomy & Physiology 50:111-130.

Tobias, Phillip V. 1968. Homage to Emeritus Professor Raymond Dart on his 75th birthday, 4th February 1968. South African Journal of Science 64:41-50.

Tobias, Phillip V. 1971. The Brain in Hominid Evolution. New York: Columbia University Press.

Tobias, Phillip V. 1973a. New developments in hominid paleontology in south and east Africa. Annual Review of Anthropology 2:311-334.

Tobias, Phillip V. 1973b. Implications of the new age estimates of the early South African hominids. Nature 246:79-83.

Tobias, Phillip V. 1974. Taxonomy of the Taung skull. Nature 252:85-86.

Vallois, H. V. 1954. La capacité cranienne chez les primates superieurs et le "Rubicon cerebral." Comptes Rendus Hebdomadaires des Seances de l'Academie des Sciences, Paris 238:1349-1351.

van Riet Lowe, C. 1948. Cave breccias in the Makapan Valley. In The Robert Broom Commemorative Volume. Alex. L. Du Toit, ed. Pp. 127-132. Cape Town: Royal Society of South Africa.

Washburn, S. L. 1950. The analysis of primate evolution with particular reference to the origin of man. Cold Spring Harbor Symposia on Quantitative Biology 15:67-78.

Weidenreich, Franz. 1944. Giant early man from Java and south China. Science 99:479-482.

Weidenreich, Franz. 1946. Apes, Giants and Man. Chicago: University of Chicago Press.

Weinert, Hans. 1950. Über die neuen Vor- und Frühmenschenfunde aus Afrika, Java, China und Frankreich. Zeitschrift für Morphologie und Anthropologie 42:113-148.

Woodward, Arthur Smith. 1925. The fossil anthropoid ape from Taungs. Nature 115:235-236.

Yates, F. and M.J.R. Healy. 1951. Statistical methods in anthropology. Nature 168:116-117.

Zuckerman, S. 1928. Age-changes in the chimpanzee, with special reference to growth of brain, eruption of teeth, and estimation of age; with a note on the Taungs ape. Proceedings of the Zoological Society of London 1928:1-42.

Zuckerman, S. 1950a. South African fossil anthropoids. Nature 165:652.

Zuckerman, S. 1950b. South African fossil hominoids. Nature 166:158-159.

Zuckerman, S. 1950c. South African fossil hominoids. Nature 166:953-954.

Zuckerman, S. 1950d. Taxonomy and human evolution. Biological Reviews 25:435-485.

Zuckerman, S. 1951a. Comments on the dentition of the fossil australopithecines. Man 51:20.

Zuckerman, S. 1951b. The dentition of the Australopithecinae. Man 51:32.

Zuckerman, S. 1952. An ape of the ape? Journal of the Royal Anthropological Institute of Great Britain and Ireland 81:57-68.

Zuckerman, S. 1954. Correlation of change in evolution of higher primates. In *Evolution as a Process*. Julian Huxley, A. C. Hardy and E. B. Ford, eds. Pp. 300-352. London: George Allen & Unwin Ltd.

TURNING POINTS IN EARLY HOMINID STUDIES

Becky A. Sigmon
University of Toronto

Introduction

The growth of the discipline of human palaeontology seems to be strongly related to the level of development present at any given time in the surrounding intellectual environment. For example, when a major set of new hominid fossil data is discovered, one can expect to see a series of patterned events unfold. These events begin with the presentation of the descriptive aspects of the fossil data and their preliminary interpretation. The next stage is seen to unfold with the scientific response which is frequently doubting and dubious. Questioning and/or rejection of initial interpretations follow with subsequent reanalyses and reinterpretations. After a period of time the descriptive results will be fitted into the overall scheme of human evolutionary events as they are known at that time. Interpretations on the systematics of the finds will be argued until major schools of thought emerge and envelope varying interpretations under their rubric.

Each major set of finds serves as a turning point in the growth of the discipline. These turning points enable us to reconsider past finds in terms of the newly made discoveries, as well as vice versa. Each new turning point effects a re-evaluation of all previous ones. In this way the field grows and matures in its understanding of events in human evolution, but it does so always with respect to its innate limitations.

Because we are sampling such a minute fraction of life forms - usually only a fragment of a bone of an individual of a population - we will never be able to see more than a glimpse of our evolutionary past. At best we are only fitting pieces of time and space together to produce a vague collage of past events. This innate limitation in the study of fossil life forms leads to the inevitable conclusion that problems of interpretation will arise among different observers. The problems are particularly manifested in the area of systematics (gaps in the fossil record will always obscure relationships) and in the reconstruction of patterns of behavior. We are in reality dealing not with individuals or populations, rather we are working with a single aspect of an individual - the skeletal framework - and even then only with fragments of this single aspect of the individual.

Considering these limitations that are innate in the field, as well as the pattern of response that usually occurs after the release of information concerning new hominid fossil finds, it is remarkable that the discipline has progressed as far as it has in understanding the events of human evolution.

Background: Early Hominid Discoveries

Dart's publication in 1925 on the discovery of the first "australopithecine" was a major landmark in human palaeontology. A new phase of evolutionary studies should have been initiated as a result, but at the time of his announcement of the Taung child's skull, the scientific community was still mentally groping with acceptance of another idea. The primitive fossils found in Java and China (later to be known as *Homo erectus*) were just beginning to be accepted as representatives of a pre-*Homo sapiens* stage of human evolution. To be confronted so soon with a possible pre-*Homo erectus* ancestor, when the former was still such a newly accepted member of the family, was too much, too soon, for most of the scientific world. The Taung child was therefore generally regarded as too simian to be ancestral to Man. The combination of hominine and simian features on the Taung skull was skeptically explained as a characteristic typical of immature material: the young of species tend to have more traits in common with the young of related species than do the adults of those species. Had the skull been that of an adult, it was felt, than there would have been little doubt as to its simian affinities.

Some scientists were less skeptical than others. Broom, for example, visited Dart to see the fossil. His examination of it led him to concur with Dart's conclusion. Broom's interest in the implications of the fossil led him to search for further evidence to document the existence of an "australopithecine" stage of human evolution. Success came in 1936 with the discovery at Sterkfontein of parts of an adult skull including a braincast (Broom 1936, 1951). There could be no doubt that this cranium from a fully mature individual was more hominine than simian, and that it represented evidence for the presence of a pre-*Homo erectus* stage of human evolution. From this point on, the study of the "australopithecines" became a seriously recognized area of scientific investigation.

South Africa became the center of "australopithecine" studies. Large samples of these early hominids were excavated from limestone quarry sites. Nearly all the evidence of this early stage of human evolution (until the Olduvai discovery in 1959) was uncovered in South Africa. Some important exceptions were the fossils found in East Africa by Louis Leakey (1932, 1933) and Kohl-Larsen (1943). The finds made by Leakey convinced the scientific community that Man had evolved at an earlier date than was previously suspected. The later multitude of fossil discoveries from South Africa substantiated this finding.

During the 1950's, taxonomic schemes and other evolutionary interpretations were published on studies of the South African material. The scientific community again balked at some of the interpretations, and counter-hypotheses were suggested. Even now, a review of the literature reveals the lack of concurrence among anthropologists as to what to call these early hominids, and how to interpret their role in human evolution. The fact remains, however, that our first knowledge of this early stage of human evolution was based on the discoveries from South Africa. The interpretations made of these fossils have influenced our thinking with respect to the evolutionary events that occurred during this early human time period. From that time on palaeoanthropologists making analyses of further finds would be constrained to deal with both the taxonomy and the evolutionary hypotheses proposed for the South African fossils.

New discoveries outside of South Africa were first made at Olduvai Gorge. "Zinjanthropus" was described by L. Leakey at 1959 and Homo "habilis" was brought to light soon afterward (L. Leakey et al. 1964). These finds received immediate international attention; the welcome from the academic community was in contradistinction to the response made when the first South African discovery was announced. The intervening 34 years of study of the South African fossils had paved the road and the scientific community was better prepared to accept the new fossils and their implications. Even the lay public, most of whom had never heard of the South African "australopithecines," became aware of the important discoveries at Olduvai Gorge. But to the scientific community, the major results of these discoveries meant firstly, that absolute dates could be determined to age the fossils (the South African fossils both then and now could be dated only by relative, not absolute means) and secondly, that additional samples of early hominids from a different geographical region were now available to make comparative studies possible.

At the same time new problems were created. The questions arose as to whether or not the Olduvai fossils were sufficiently distinct from the South African ones to merit separate taxonomic designation. What patterns did the Olduvai and the South African fossils have in common; how did they differ? Where in the evolutionary lineage of Man did the Olduvai fossils belong?

Before these newly emerging problems could be studied in detail, the search for early hominids began in Africa's Eastern Rift Valley. Multidisciplinary studies, including palaeoanthropological research, were set up to investigate various aspects of the palaeoenvironment of late Pliocene-early Pleistocene times. Large scale palaeontological expeditions were mounted by Howell, Arambourg, Coppens and R. Leakey in 1967 in the Lower Omo Valley, Ethiopia (Howell 1968), by R. Leakey in 1968 on the eastern shores of Lake Turkana, Kenya (R. Leakey, 1970), and by Taieb and Johanson in 1972 in the Central Afar, Ethiopia (Taieb, et al.,1972,1974). These multi-faceted studies have been continued into the present decade and have resulted in the discovery of yet additional early hominid fossils, environmental data, and absolutely dated deposits. This additional information posed further questions regarding nomenclature and interpretation of human evolutionary stages.

To date, the search for remains of early hominids has produced reasonably large samples of these fossils, (for summary see Tobias 1972, 1973, 1974; Sigmon 1977), a great deal of associated environmental information, data on floral and faunal remains, and relative and absolute chronological sequences. At present it will require many years for the descriptive studies of this information to be completed. In the meantime there remain a number of unresolved problems which have arisen as a result of the past half century of research.

Problems which have been created: Nomenclature

Presently there are several taxa which have been proposed following the rules of the International Code of Zoological Nomenclature (Stoll 1964). These are _Homo africanus_ and _Paranthropus robustus_ (Robinson 1954, 1972)

H. habilis (Leakey, Tobias and Napier 1964), and *H. ergaster* (Groves and Mazak 1975). Of these schemes, it is my opinion that *H. africanus* and *Paranthropus robustus* have the most substantial documentation as valid taxa. The case for *H. habilis* and *H. ergaster* seem a great deal weaker.

Other taxonomic usages exist as *suggested* schemes, but of the following none has been submitted as a formal proposition: Mayr (1950) refers to all early hominids as *Homo transvaalensis*; Le Gros Clark (1955) "provisionally" recognized one genus of early hominid which he referred to as *Australopithecus*; Campbell (1963) suggested that the South African hominids be referred to as one genus and two species, *Australopithecus africanus* and *A. robustus*. Most students would agree that Broom's original taxonomic scheme (Broom 1950) of three subfamily and five generic names is outdated and not applicable to the fossil record as we now know it. Le Gros Clark's and Campbell's schemes have acquired much popularity, especially by authors of introductory textbooks. The latter also tend to be especially fond of using the indecisive terms the "robustus australopithecine" in reference to *Paranthropus*, and the "gracile australopithecine" to indicate *Homo africanus*.

Evidently we are far from universal agreement in the naming of early hominids. It was not unpredictable that finds made outside of South Africa would set off further nominal confusion. In 1959 at Olduvai Gorge a robust hominid skull was found and names "Zinjanthropus" (L. Leakey 1959). Although Leakey followed the customary procedure of comparing the new find with previously described fossils resembling it, namely *Australopithecus* (=*Homo*) and *Paranthropus*, he did not deem the similarities sufficient at that time to place the new fossil into either taxon. Later, he regarded "Zinjanthropus," *Paranthropus* and *Australopithecus* as subgenera of the genus *Australopithecus* (Leakey, Napier and Tobias 1964). That "Zinjanthropus" was not another new hominid genus or subgenus was demonstrated by later studies which showed that it shared the basic adaptive pattern present in *Paranthropus*. "Zinjanthropus" was subsequently described in a monograph entitled *Australopithecus (Zinjanthropus) boisei* (Tobias 1967); although this notation indicated subgeneric status for "Zinjanthropus," Tobias himself proposed that the nomen "Zinjanthropus" be sunk.

A similar problem exists with the taxon *Homo* "*habilis*" (Leakey, Tobias, and Napier 1964) which several workers consider to be an unconfirmed or invalid taxon (Oakley & Campbell 1964; Robinson 1965, 1966; Simons and Pilbeam 1965; Bielicki 1966; Simons 1968; Brace et al. 1973). It has been suggested that the fossils categorized as Homo "habilis" consist of two separate taxa, namely Homo africanus and Homo erectus (Robinson 1965; Simons et al. 1969; Brace et al. 1973). When these fossils are described in a future monograph, these problems will surface, and the solutions should be revealing. As a comparative sample, Homo "habilis" could illuminate certain murky waters of present knowledge.

Recently a new taxon, Homo ergaster, was created by Groves and Mazak (1975) based almost entirely on dentition. At present it would seem judicious to exercise restraint in regarding this as a new taxon until it can be more clearly and fully demonstrated that it possesses a morphological pattern different from that of other species previously proposed.

Naming of hominids from the Eastern Rift Valley has been tentative and suggestive. There are apparently at least two, or possibly three, lineages occurring there, one (or two) with affinities to Homo and the other with affinities to Paranthropus. The naming of the populations represented is an unsettled and confusing issue because of the differing authors. For example, the genus "Australopithecus" has been used generally by R. Leakey (1971, 1972) to indicate the East African form of Paranthropus. More recently he has suggested that the "gracile australopithecine" Australopithecus cf. africanus may also be present there; thus it becomes necessary to differentiate the latter from the former taxon which he feels most closely resembles A. "boisei" (R. Leakey 1976a). Two forms of Homo are recognized, one occurring at early levels and resembling H. "habilis" (Leakey 1976a); the other occurs later in the sequence at 1.6-1.3 million years and resembles H. erectus (R. Leakey and Walker 1976). A. "boisei" occurs throughout the sequence. If one follows the suggestion of Robinson (1965) and others that H. "habilis" is an advanced variant of H. africanus, then we are left with three lineages at East Rudolf. They are Paranthropus, Homo africanus and Homo erectus. The former two taxa occur throughout the sequence at about 2.6* (Fitch and Miller 1970, 1976) or 1.85 million years (Curtis 1975; Curtis et al. 1975) to about 1.3 million years ago. Homo erectus appears later from about 1.6 to 1.3, but overlaps in time and space with the other two lineages.

*See Addendum.

The Omo fossil sample has been described as containing members which resemble *Australopithecus robustus* or *A. boisei*, *A. africanus* and Bed I *Homo* "*habilis*" (Howell 1968, 1969, 1972), with dates ranging from 2+ for the former and 2-3 million years for the latter two.

At Hadar, fossils dated at about 3 million years have been described, some of which are said to bear resemblances to *A. africanus*, others to *A. robustus*, and others which are called *Homo sp. indet*. (Taieb et al. 1976; Johanson and Taieb 1976).

The following chart attempts to compare nomina commonly used (left) with the scheme (minus *Homo sp. indet*) proposed by Robinson (1972) (right):

Locality		
Olduvai Gorge	*Homo habilis*	*Homo africanus* and *Homo erectus*
	Australopithecus boisei	*Paranthropus*
Lower Omo Valley	*A. africanus*	*H. africanus*
	A. robustus or *A. boisei*	*Paranthropus*
East Rudolf	*A. cf. africanus*	*H. africanus*
	A. boisei	*Paranthropus*
	H. cf. habilis	*H. africanus*
	H. erectus	*H. erectus*
Hadar	*A. africanus*	*H. africanus*
	A. robustus	*Paranthropus*
	Homo sp. indet.	*Homo sp. indet*.

One point seems clear. Two lineages of hominids co-existed (as had been previously suggested by South African data) in East Africa from at least 3 million years to about 1 million years ago. One lineage, *Paranthropus*, remained relatively unchanged throughout this 2 million year period. The other or *Homo* lineage is less clearly definable at present.

Most workers will agree that the *Homo* lineage includes *Homo sapiens* and *Homo erectus*, but events that occurred previous to the evolution of these taxa is still a clouded issue. Again the problem arises as to where *Homo* "*habilis*" fits. This issue becomes re-emphasized with the discovery of a new fossil from

Sterkfontein which is said to have affinities with Homo "habilis", and which occurs in deposits more recent than Homo africanus (Hughes and Tobias 1977). And the argument reappears as to the taxonomic status of Homo africanus and whether its representatives merit being recognized as members of the genus Homo. If indeed at East Rudolf and Hadar three lineages exist, all at about the same time, the possibility and the problem is raised of two co-existing lineages of Homo, one a more advanced form like Homo erectus and one a less advanced form like Homo africanus. The probable presence of a similar kind of situation (at least to that of East Rudolf) has been suggested as occurring also in South Africa (Robinson 1961). Such an occurrence might be explained, he wrote, by the migration of the more advanced toolmaking Homo erectus into the area inhabited by Paranthropus and Homo africanus, resulting in the ultimate displacement of the competitive but less advanced Homo africanus. At present the frequent use of the term Homo sp. indet. is a good indication of our present unsettled state of knowledge.

A notable limitation in the further study of any and all new finds, especially those being recovered from the Eastern Rift Valley, is the present incompleteness in the study of earlier discovered fossils. Although there is a great deal of information published on the South African fossil hominids in the form of articles and monographs, the letter form is usually the more essential for comparative analyses. A monographic treatment of the cranial material remains to be completed (presently in preparation by J.T. Robinson). Otherwise most of the pre-1965 South African finds have been described and the analyses occur in a series of monographs published by the Transvaal Museum and elsewhere (e.g. Broom and Schepers 1946; Broom, Robinson and Schepers 1950; Broom and Robinson 1952; Robinson 1956; Dart 1957; Brain 1958; Robinson 1972). The date of 1965 is used here because it is the year that Brain reopened excavations at Swartkrans; the following year Tobias re-established excavations at Sterkfontein.

Of the Olduvai Gorge fossil hominids, "Zinjanthropus" has received detailed monographic treatment (Tobias 1967), but a total descriptive analysis of the Homo "habilis" finds has not yet been undertaken.

These lacunae in descriptions of early finds remain a distinct problem in analyzing new finds from South and

East Africa, but especially for those being recovered in the Eastern Rift Valley. It also provides one explanation for the less than discriminating use of nomina.

Other Problems: Interpretation

Evolutionary interpretations that were made based on the early discoveries of fossil hominids from South Africa have strongly affected the study of human palaeontology. An enumeration of major points deduced from the fossil evidence there begins first with a confirmation of the belief suggested by previous *Homo erectus* finds that erect bipedalism evolved before the brain began to increase in size. Secondly, the fossil data suggested that the possession of erect posture did not in itself guarantee the "evolving" of a hominid to a hominine form. *Paranthropus* is an example of a hominid that walked upright but did not evolve a progressively larger brain and probably had no more than a rudimentary culture, if that. Rather, the evidence reveals that it remained a relatively stable, conservative lineage through at least 2 million years (these dates are based on East African finds) and then became extinct. Thirdly, the South African material was the basis for the proposal of the simultaneous occurrence of two sympatric hominid taxa (Robinson (1954). Fourthly, it formed the evidence used to construct the "dietary" hypothesis that suggested that *Paranthropus* and *Australopithecus* (=*Homo*) had significantly different dietary patterns (Robinson 1961, 1963). And finally, from this material sprang the hypothesis that behavioral differences provided a major explanation for the morphological differences in the two lineages. *Homo africanus*, it was suggested, had undertaken a new means of adapting to the environment by the use of culture, whereas *Paranthropus* had not, to any significant degree, employed cultural means of adapting (Robinson 1964, 1968).

These contributions to the study of human evolution have served as models, some of which would be confirmed in fact by later studies, while others would lead to continued questioning and expanded or different interpretations. The interpretations based on the South African discoveries were to have long term effects on subsequent studies; students would have to consider their validity or demonstrate their invalidity before proceeding with further analyses.

For example, although the presence of erect bipedality in the South African hominids was universally agreed upon, its level of development was not. Three basic points of view demonstrate this lack of concurrence. One interpretation suggests that although the two taxa were both erect bipeds, they were differently adapted, Homo africanus being adapted as efficiently as modern Man, but Paranthropus being less efficient in comparison with modern Man and with tendencies inclining it toward certain aspects of a pongid pattern of locomotion (Robinson 1972). To others (e.g. Zihlman 1970; Zihlman and Hunter 1972; Lovejoy 1973, 1975; Lovejoy et al. 1973; McHenry 1975a, 1975b) the evidence does not reveal differences in patterns of bipedality for the two taxa but rather that both possess a fully developed erect bipedal gait which was antecedent to modern Man. Yet a third interpretation (Oxnard 1975a, 1975b) suggests that the type of locomotion and posture present in both South African taxa was not sufficiently developed to be ancestral to modern Man. In these three instances the authors are using essentially the same evidence yet are reaching different conclusions. The question comes to mind as to how much these differences are a reflection of initial interpretation of the evolutionary position of these taxa.

The early development of erect posture is further substantiated by discoveries at East Rudolf and Hadar. The Ethiopian finds push back its occurrence to 3 million years and the East Rudolf fossils extend it nearly that far. Further detailed analyses of this material will most probably provide additional clues about the level of development and variation of erect bipedalism among different hominids.

Another question which has arisen from the models is that of the ancestral form which gave rise to both Paranthropus and Homo. The evidence from South and East Africa indicate conservative rates of evolution for Paranthropus, but fast rates for Homo. The fossil record thus far suggests the earliest presence of Paranthropus at about three million years (Hadar), and for Homo (?) possibly at about five and a half million years (Lothagam) (Patterson, et al. 1970). The occurrence of any hominid finds that are earlier than three million years is extremely rare. The earliest group of fossils considered to be hominids, and consisting of some jaw fragments and teeth, are those from Laetolil, representing at most an age of about 3.7 million years (M.Leakey et al. 1976; White 1977).

The fossil record is too incomplete to justify a conclusion as to which lineage most resembled the common ancestor. Both taxa have been considered for this position (e.g. Tobias 1967; Robinson 1972). The question can only be solved by filling in gaps in the fossil record and by reconstructing the selective forces that would have been operating at the time of divergence. Some of the latter have been suggested (Brian 1958; Robinson 1963). In addition, theoretical speculation based on our understanding of the process of evolution can direct our thinking, but not solve the issue. It is usually the case that the more generalized, conservative forms most closely resemble the ancestral one, and that the more rapidly evolving and occasionally more specialized forms are less like the "primitive" antecedent. This argument would suggest that the rapidly evolving *Homo* lineage sprang from the more conservative and slowly evolving lineage of *Paranthropus*. However, the fossil record at present contradicts this theoretical argument because the earliest hominids thus found appear to be more hominine than paranthropine. The question arises as to which of the two lineages is actually the more specialized? It is easy to see that one's view of the taxonomic distinctions between these lineages could affect one's interpretation.

With respect to other points of interpretation, the finds at Olduvai are interesting in that they bring into clearer focus the issue of the occurrence of sympatric hominid lineages. The Olduvai finds provided further support for the two lineage hypothesis. At the same time they raised the question of the evolutionary affinities between the two lineages present at Olduvai on the one hand, and the two lineages in South Africa on the other hand (see discussion on "*Zinjanthropus*" and *Homo* "*habilis*"). The Olduvai discoveries confirmed the association between culture and the hominids, but they have not resolved the question of which form possessed the cultural adaptation. As in South Africa, the deduction was made that the form morphologically more like modern Man, i.e. that belonged to the *Homo* lineage, was the more likely candidate for toolmaker.

Research in the Eastern Rift Valley also supported the existence of at least two sympatric lineages - a *Homo* and a non-*Homo* lineage (Howell 1968, 1969; Leakey 1976b; Johanson and Taieb 1976). But again the question arose as to which hominid was culture bearing.

More tools and tool making activity were uncovered (Isaac 1976) and at increasingly earlier dates, but evidence for direct association between artifacts and hominid makers has continued to be elusive. Still the reasonable assumption is made that the tools were fashioned by members of the *Homo* lineage.

The fossil record from the Rift Valley is continuing to present the field with new problems and unique discoveries. It raises the particularly tantalizing question of the composition of the *Homo* lineage. Could two varieties of hominines have co-existed - a variety with a smaller brain and perhaps less well developed culture, with a form that had a slightly larger brain, better developed culture, and somewhat different dental pattern? Both types occur at early time horizons. How long could they have co-existed, living side by side? Or are we testing samples of geographical variants which did not normally overlap in space? How can the variations seen in these advanced *Homo* of East Africa along with the suggestion of different evolutionary patterns and perhaps faster evolutionary activity be reconciled with the evolutionary pattern of the South African and at least some of the Oldvai fossils? These are major questions to be answered, hopefully, in the near future; the answers lie in years of study of this newly emerging set of data.

The Rift Valley studies are also in the process of producing large amounts of information on palaeoenvironment. Answers are beginning to emerge concerning the patterns of living behavior of early hominids, and these in turn may provide answers to one of the least known aspects of the two (or more?) evolutionary lineages. That is, how did the lineages differ ecologically? The division of taxa into separate adaptive plateaux implies the presence of distinct differences of morphology, behavior and ecological exploitation. If additional evidence can be produced for the latter, then we are a step further along the way of understanding our ancestral evolutionary patperns.

In Conclusion

For over fifty years human palaeontologists have been trying to unravel the unknowns regarding the stages of human evolution that preceded *Homo erectus*. Much progress has been made in interpreting the fossil record of this time period. Still, certain primary questions remain unresolved, such as how did the early hominids adapt to their environment, and what are the evolutionary relationships among the different groups of hominids now known.

The hypotheses that were formulated based on the South African discoveries have influenced our thoughts with respect to these questions. The discoveries that were subsequently made in East Africa greatly complicated the picture. They introduced new data from the fossil record and accompanying this were new problems. As a result we are presently trying to sort out the evolutionary relationship between the hominid populations from South and East Africa, as well as trying to sort out the meaning of the differences seen in hominids who lived coevally in the same areas. In a sense we seem to have reached the end of one phase of study and are turning our thoughts to another as more and more of our evolutionary background is brought to light. One might say that we are at the beginning of a new half century of probing into our evolutionary past.

Addendum

This paper was written four years ago and delay in publication allows me the opportunity to discuss recent advances made in the field of Palaeoanthropology. One issue that has apparantly been settled is the dating of the KBS tuff at East Turkana. Fossils previously thought to be 2.6 million years in age actually fall into a younger age category. Fission track dating by Gleadow (1980) and K-Ar dating by McDougall et. al. (1980) have fixed the basal date of this tuff as about 1.87 to 1.89 million years (see page 6 of text).

A second major addition to human paleontological research is the Hadar sample of fossil hominids from the Afar Triangle of Ethiopia which "...constitute the earliest substantial [my italics] record of the family Hominidae." (Johanson and White 1979; 321) This sample is extraordinary because of the number of excellent preserved specimens recovered, the possibility it provides in reconstructing individuals, and the likelihood that one site is sampling a population or perhaps even a family of related individuals. In addition, absolute and relative dates are retrievable as volcanic products are prevalent and associative faunal remains are abundant in the sediments. Studies on the Hadar material is greatly enlarging our knowledge of human evolution and ranges of variation within the hominid lineage during the Pliocene.

Where does the Hadar material fit in the classificatory scheme of hominid evolution? Not unexpectedly, this issue is presently controversial. The Hadar fossils have been combined with those from Laetoli and a new taxon has been proposed, *Australopithecus afarensis* (Johanson, White and Coppens 1978). Strictly speaking, the proposal of this taxon is at present incomplete. Firstly, the proposers base their definition of the generic term *Australopithecus* on the "provisional" defition suggested by Le Gros Clark (1955). The Code of Zoological Nomenclature does not recognize provisional proposals. Secondly, it is generally understood that setting up a new taxon requires firstly a comparison with closely related types in order to demonstrate the distinctiveness of the new taxon, and secondly the provision of statements that support the validity of the taxon, i.e. that show its uniqueness from other existing taxa. With regard to these requirements, the proposers have omitted consideration of other taxonomic proposals and have relied on a proposal "provisionally" suggested in 1955. In addition, subsequent finds and taxonomic interpretations are not considered in this newly proposed taxon. Tobias (1978, 1980), for example, has objected to this proposal because he feels that the Hadar/Laetoli sample overlaps in time and morphology with that of *Australopithecus africanus*. He draws attention especially to the more recent evidence that suggests earlier dates for the South African sites than had previously been considered. Late Pliocene dates of around 3 million years for Member 3 at Makapansgat and Member 4 at Sterkfontein suggest roughly contemporary time ranges with Hadar.

Further analysis and interpretation in a subsequent publication by Johanson and White (1979) on hominid systematics provides limited comparisons with other related fossil material, but the same format as the earlier paper is employed. If a new taxon is indeed required, as it may be for this sample, the authors have not yet sufficiently documented it.

The revised Hadar interpretation (Johanson, White and Coppens 1978; Johanson and White 1979) indicates the absence of *Paranthropus* at that site and thus alters the chronology of this taxon, reducing its known time range back to a little over 2 million years to a little less than 1 million years ago. The Hadar material sheds no further light on the evolutionary affinities of *Paranthropus*.

In conclusion, the past several years have been exceedingly rich in providing the field with increasingly larger samples of fossil hominids. Despite the inconclusive taxonomic distinctions, the descriptive aspects of the study of human evolution are broadening. In a field that is notoriously deficient in data, it is phenomenal - and a tribute to the scientists - that our understanding of the evolutionary stages and the ranges of variation in early hominid taxa is as advanced as it actually is.

Acknowledgments

I gratefully acknowledge the following people who, throughout the last several years, have generously allowed me to examine early hominid material in their possession: C.K. Brain, F.C. Howell, D.C. Johanson, L.S.B. Leakey, M. Leakey, R.E.F. Leakey, J.T. Robinson, M. Taieb and P.V. Tobias. Financial aid for helping to meet travel costs has been provided by the University of Toronto. My appreciation also is to be noted to P.L. Storck who kindly read and made suggestions on the manuscript. I alone assume responsibility for the interpretations.

References

Bielicki, T. 1966. On "Homo habilis." Cur. Anthrop. 7:576-578.

Brace, C-L. P.E. Mahler, and R.B. Rosen. 1973. Tooth measurements and the rejection of the taxon "Homo habilis." Yearbk. Phys. Anthrop. 16:50-68.

Brain, C.K. 1958. The Transvaal Ape-Man Bearing Cave Deposits. Transvaal Museum Memoir No. 11.

Broom, R. 1936. A new fossil anthropoid skull from South Africa. Nature 138:486-488.

_____. 1950. The genera and species of the South African Fossil Apemen. Am. J. Phys. Anthrop. 8:1-13.

_____. 1951. Finding the Missing Link. Watts and Co., London.

Broom, R., J.T. Robinson, and G.W.H. Schepers. 1950. Further Evidence of the Structure of the Sterkfontein Ape-Man Plesianthropus. Transvaal Museum Memoir No. 4. Transvaal Museum, Pretoria.

Broom, R. and G.W.H. Schepers. 1946. The South African Fossil Ape-Men. The Australopithecinae. Part I. The Occurrence and General Structure of the South African Ape-Men. Transvaal Museum Memoir No. 2:7-144.

Broom, R. and J.T. Robinson. 1952. Swartkrans Ape-Man. Paranthropus crassidens. Transvaal Museum Memoir, Pretoria, No. 6.

Campbell, B. 1963. "Quantitive taxonomy and human evolution," in Classification and Human Evolution. Edited by S.L. Washburn. Viking Fund Publ. in Anthrop. No. 37. New York: Wenner-Gren Foundation for Anthropological Research, Inc.

Curtis, G.H. 1975. Improvements in potassium-argon dating: 1963-1975. World Arch. 7(2):198-209.

Curtis, G.H., Drake, T. Cerling and Hampel. 1975. Age of KBS Tuff in Koobi Fora Formation, East Rudolf, Kenya. Nature 258:395-398.

Dart, R. 1925. Australopithecus africanus: the Man Ape of South Africa. Nature (115:195-199.

_____. 1957. The Osteodontokeratic Culture of Australopithecus prometheus. Transvaal Museum Memoir No. 10. Transvaal Museum, Pretoria.

Fitch, F.J. and J.A. Miller. 1970. Radioisotopic age determinations of Lake Rudolf artefact site. Nature 226:226-228.

_____.1976. Conventional potassium-argon and argon-40/argon-39 dating of volcanic rocks from East Rudolf. Earliest Man and Environments in the Lake Rudolf Basin: Stratigraphy, Palaeoecology and Evolution, ed. Y. Coppen, F.C. Howell, G.L. Isac, and R.E.F. Leakey, pp. 123-147. University of Chicago Press, Chicago.

Groves, C. and V. Mazak. 1975. An approach to the taxonomy of the Hominidae: gracile Villafranchian hominid of Africa. Casopis pro mineralogii a geologii 20(3):225-247.

Howell, F.C. 1968. Omo research expedition. Nature 219:567-572.

_____.1969. Remains of Hominidae from Pliocene/Pleistocene formations in the Lower Omo Basin, Ethiopia. Nature 223:1234-1239.

_____. 1972. "Pliocene/Pleistocene Hominidae in Eastern Africa: absolute and relative ages," in Calibration of Hominoid Evolution. Edited by W.W. Bishop and J.A. Miller. Edinburgh: R. and R. Clarke Ltd.

Hughes, A. and P.V. Tobias. 1977. A fossil skull probably of the genus Homo from Sterkfontein, Transvaal. Nature 265 (5592):310-312.

Isaac, G.L. 1976. Plio-Pleistocene artifact assemblages from East Rudolf, Kenya. In Earliest Man and Environments in the Lake Rudolf Basin, ed. Y. Coppen, F.C. Howell, G.L. Isaac, and R.E.F. Leakey, pp. 552-564, University of Chicago Press.

Johanson, D.C. and M. Taieb. 1976. Plio-Pleistocene Hominid discoveries in Hadar, Ethiopia. Nature 260(5549):293-297.

Kohl-Larsen, L. 1943. Aufden Spuren des Vormenschen. Forschungen, Fahrten und Erkentuisse in Deutsch-Ostafrika (Deutsch Afrika-Expedition 1934-1936 und 1937-1939). 2 vols. Stuttgart: Strecker und Schroder Verlag.

Leakey, L.S.B. 1932. The Oldoway human skeleton. Nature 129:721-722.

_____.1933. The status of the Kanam mandible and the Kanjera skulls. Man 33:200-201.

_____.1959. A new fossil skull from Olduvai. Nature 184:491-493.

Leakey, L.S.B., P.V. Tobias, and J.R. Napier. 1964. A new species of the genus Homo from Olduvai Gorge. Nature 202:7-9.

Leakey, M.D., R.H. Hay, G.H. Curtis, R.E. Drake, M.K. Jackes, and T.D. White. 1976. Fossil hominids from the Laetolil Beds. Nature 262:460-466.

Leakey, R.E.F. 1970. New hominid remains and early artifacts from Northern Kenya. Nature 226:223-224.

_____. 1971. Further evidence of Lower Pleistocene hominids from East Rudolf, North Kenya. Nature 231:241-245.

_____.1972. New fossil evidence for the evolution of Man. Soc. Biol. 19:99-114.

Leakey, R.E.F. 1976a. New hominid fossils from the Koobi Fora formation in Northern Kenya. Nature, 261(5561):574-576.

_____.1976b. Hominids in Africa. Amer. Scient.64(3): 174-178.

Leakey, R.E.F. and A.C. Walker. 1976. Australopithecus, Homo erectus and single species hypothesis. Nature 261 (5561) 572-574.

Le Gros Clark, W.E. 1955. The Fossil Evidence for Human Evolution. University of Chicago.

Lovejoy, C.O. 1973. The gait of Australopithecines. Yrbk. Phys. Anthrop. 17:147-161.

_____.1975. Biomechanical perspectives on the lower limb of early hominids. In Primate Functional Morphology and Evolution, ed. R.H. Tuttle, pp. 291-326, Mouton Publishers, The Hague.

Lovejoy, C.O., G.H. Heiple and A.H. Burstein. 1973. The gait of Australopithecus. Am. J. Phys. Anthrop. 38 (3):757-780.

Mayr, E. 1950. Taxonomic categories in fossil hominids. C.S.H.S. Quant. Biol. 25:109-118.

McHenry,H. 1975a. Biomechanical interpretation of the early hominid hip. J. Hum. Evo. 4:343-355.

_____. 1975b. The ischium and hip extensor mechanism in human evolution. Am. J. Phys. Anthrop. 43:39-46.

Oakley, K. and B. Campbell. 1964. Newly described Olduvai hominid. Nature 202:732.

Oxnard, C.E. 1975a. The place of the australopithecines in human evolution: grounds for doubt? Nature 258:389-395.

_____.1975b. Uniqueness and Diversity in Human Evolution: Morphometric Studies of Australopithecines. University of Chicago Press.

Patterson, B., A.K. Behrensmeyer, and W.D. Sill. 1970. Geology and fauna of a new Pliocene locality in northwestern Kenya. Nature 226:918-921.

Robinson, J.T. 1954. The genera and species of the Australopithecinae. Am. J. Phys. Anthrop. 12:181-200.

_____. 1956. The Dentition of the Australopithecinae. Transvaal Museum Memoir No. 9, Transvaal Museum, Pretoria.

Robinson, J.T. 1961. The Australopithecines and their bearing on the origin of Man of stone tool-making. S. Afr. J. of Sci. 57:3-16.

_____. 1963. "Adaptive radiation in the australopithecines and the origin of Man," in African Ecology and Human Evolution. Edited by F.C. Howell and F. Bourliere. Viking Fund Publ. in Anthrop. No. 36. New York: Wenner-Gren Foundation for Anthropological Research Inc.

______. 1964. Some Critical phases in the evolution of Man. Arch. Bull. 19:3-12.

______.1965. Homo 'habilis' and the australopithecines. Nature 205:131-134.

______.1966. The distinctiveness of Homo "habilis." Nature 204:957-960.

______. 1968. The origin and adaptive radiation of the australopithecines. In Evolution and Hominisation, ed. G. Keuth, 150-175, Gustav Fischer Verlag, Stuttgart.

______. 1972. Early Hominid Posture and Locomotion. Chicago: University of Chicago Press.

Sigmon, B.A. 1977. Contributions from Southern and Eastern Africa to the Study of early hominid evolution. J. Hum. Evo. 6(3):245-257.

Simons, E.L. 1968. Assessment of a fossil hominid. Science 160 (3828): 672-675.

Simons, E.L. and D.R. Pilbeam. 1965. Some problems of hominid classification. Amer. Sci. 53:237-259.

Simons, E.L., D.R. Pilbeam, and P.C. Ettel. 1969. Controversial taxonomy of fossil hominids. Science 166 (3902): 258-259.

Stoll, N.R. et. al., eds. 1964. International Code of Zoological Nomenclature. Trust for Zoological Nomenclature, London.

Taieb, M., Y. Coppens, D.C. Johanson, and J. Kalb. 1972. Dépôts sedimentaires et faunes du Plio-pleistocène de la basse vallée de l'Awash (Afar central, Ethiopie). C.R. Acad. Sc. Paris 275: 819-822.

Taieb, M., D.C. Johanson, Y. Coppens, R. Bonnefille, and J. Kalb. 1974. Découverte d'Hominidés dans le séries Plio-pleistocènes d'Hadar (Bassin de l'Awash, Afar, Ethiopie). C.R. Acad. Sc. Paris 279:735-738.

Taieb, M., D.C. Johanson, Y. Coppens, and J.L. Aronson. 1976. Geological and palaeontological background of Hadar hominid site, Afar, Ethiopia. Nature 260(5549):289-293.

Tobias, P.V. 1967. Olduvai Gorge. Vol. 2. The Cranium of Australopithecus (Zinjanthropus) boisei. Cambridge at the University Press.

_____. 1972. "Progress and problems in the study of Early Man in Sub-Saharan Africa," in The Functional and Evolutionary Biology of Primates. Edited by R. Tuttle. Chicago: Aldine.

_____.1973. New developments in hominid paleontology in South and East Africa. Ann. Rev. Anthrop. 2:311-334.

_____.1975. New African evidence on the dating and phylogeny of the Plio-Pleistocene Hominidae. Quaternary Studies, ed. R.P. Suggati and M.M. Creswell. The Royal Soc. New Zealand, 289-296.

White, T.D. 1977. New fossil hominids from Laetolil, Tanzania. Am. J. Phys. Anthrop. 46(2):197-230.

Zihlman, A.L. 1970. The question of locomotion differences in Australopithecus. Proc. 3rd. Int. Congr. Primat. 1:54-66.

Zihlman, A.L. and W.S. Hunter. 1972. A biomechanical interpretation of the pelvis of Australopithecus. Folia Primate. 18:1-19.

References for Addendum

Gleadow, A.J.W. 1980. Fission track age of the KBS tuff and associated hominid remains in northern Kenya. Nature 284:225-230.

Johanson, D.C., T.D. White and Y. Coppens. 1978. A new species of the genus Australopithecus (Primates: Hominidae) from the Pliocene of Eastern Africa. Kirtlandia No. 18:1-14.

Johanson, D.C. and T.D. White. 1979. A systematic assessment of early African hominids. Science 203:321-330.

McDougall, I., R. Maior, P. Sutherland-Hawkes, and A.J.W. Gleadow. 1980. K-Ar age estimate for the KBS tuff, East Turkana. *Nature* 284:230-234.

Tobias, P.V. 1978. Position et rôle des australopithécines dans la phylogénie humaine, avec étude particulière de *Homo* habilis et des théories contoversiés avancées à propos des premiers hominidés fossiles d'Hadar et de Laetolil. In: Origines humaines et les epoques de l'intelligence. Colloque internationale (juin 1977) de la Fondation Singer-Polignar, 38-77. Paris: Masson.

______. 1980. *Australopithecus* and early *Homo*. Proc. 8th Panafrican Cong. Prehistory and Quaternary Studies. Nairobi 1977. Ed. by R. Leakey and B. Ogat.

SEDIMENT MATRICES OF THE SOUTH AFRICAN AUSTRALOPITHECINES: A REVIEW

Karl W. Butzer
The University of Chicago

One of the most immediate records available for productive study of fossils or archeological strata is the sediment matrix that embeds the site components. This matrix commonly consists of multiple, event-specific or process-specific lenticles that document depositional micro-environments. In their turn, such micro-environments reflect the natural moisture regime, the balance of soil formation and erosion, as well as vegetation cover of the external, topographic matrix. At an even larger scale, this meso-environment forms part of a physical and biotic landscape, the regional matrix (Butzer, 1982).

The interpretation of field and laboratory analyses devoted to sequences of site deposits consequently involves a scale-translation between sedimentary, topographic and regional contexts that is or should be explicit. Clues are porvided by (1) the proportions of clay, silt and sand, which reflect sandy or clayey soils in the meso-environment as well as transport energy in the micro-environment; (2) the sorting of size-grades into distinct or blurred, and single or multiple peaks, reflecting the efficiency of one or more transport modes, so helping in their identification; (3) the arrangement of sediment into beds or laminae, subhorizontal or inclined, reflecting minute episodes of deposition and/or erosion, frequently characteristic of one particular transport mode; (4) presence or absence of cementing material, such as calcium bicarbonate, reflecting surface or subsoil water movement during or after sedimentation, in relation to specific evaporation properties within the micro- or meso-environment; and (5) erosional evidence, within or between horizons, reflecting mechanical abrasion or chemical corrosion as a result of processes such as running water, wind, or human disturbance.

The australopithecine breccias of the Sterkfontein and Makapan valleys are deep fissure or cavern fills once found beneath an undulating, karstic topography. In the Krugersdorp area, Sterkfontein proper represents the bottom of a former sink-hole (doline) below an upland surface; the nearby site of Swartkrans, on the other hand, was part of a complex doline gradually exposed on a hillside; finally, Kromdraai originally was a complex solution fissure below a bedrock platform. The other major site, Makapansgat, is found farther north, near Potgietersrus, in what once was a large double doline (uvala), then situated below the floor of a stream valley.

None of these sites was a typical cave and none was used as a hominid shelter. Instead, the fossils come from soil or alluvium washed into the caves from the outside, through near-vertical shafts or laterally, via interconnecting underground passages. Each site has several generations of deposits, reflecting intervals of external soil erosion and coarse-detrital cavern aggradation, alternating with periods of soil stability outside and limited travertine accumulation inside. Such cyclic changes between morphodynamic and morphostatic conditions continued while the external topography was reduced and dissected, gradually exposing the caverns on increasingly steep hillsides.

The individual beds of these four sites are arranged in chronological order in Table 1, as based on the faunal content (Vrba, 1976) and paleoenvironmental information. The stratigraphic units for the various site horizons or locations follow Butzer (1976) for Swartkrans, Partridge (1978) for Sterkfontein, Partridge (1979) for Makapansgat, and Brain (1981) for Kromdraai. The interpretations are all based on my own studies and do not necessarily coincide with those of other authors.

In overview, these deposits reflect a range of regional environments, at one extreme a little drier than modern Potgietersrus, on the other moderately wetter than modern Krugersdorp, i.e. between about 600 and 800 mm annual precipitation. Together with the various faunas, these results argue for a persistence of grassy environments interspersed with bush, thickets, or woodland.

The early Pleistocene Taung fossil was found in an intrusive fissure filling, originally formed within massive travertine sheets deposited by streams below a dolomite carpment (Butzer, 1974; Butzer et al., 1979; Butzer, 1980). This fill included two distinct components: (1) the older matrix of the Taung "baboon breccia" is an eolian sand, reworked by rainwash during the initial phases of sedimentation cycle, while (2) the younger, forming the matrix of

Australopithecus africanus, comprises alternating sandy silt and clayey flowstone, related to protracted spring activity and wet conditions at the height of a travertine phase. Modern travertine analogs from the eastern Transvaal suggest a mean precipitation of 600-800 mm (compared with 450mm today), inferring a vegetation of dense grass, with fringing tree growth near water sources. In other words, this australopithecine lived in a subhumid setting, much like the Transvaal australopithecines did, and not in a semidesert environment, as previously assumed.

The sum total of the sedimentary evidence indicates that the South African australopithecines inhabited a similar regional environment, with 600-800mm precipitation. Within that macro-landscape, they lived or were buried in similar topographic contexts, normally associated with stands of trees or fringing thickets within a partly-grassy or predominantly open environment. Despite a degree of meso- and macro-environmental differentiation, there is no correlation between gracile or robust australopithecines on the one hand, and drier or wetter contexts on the other. Nor is there geological support for different vegetation types or for the hypothesis of dietary differentiation between the australopithecine lineages.

Comparison of the South and East African contexts indicates that all early hominid sites were located in mosaic environments along ecotones of the seasonally, dry African "savanna" (Butzer, 1977, 1978). This argues that the Plio-Pleistocene hominids were to a large degree sympatric, and the Miocene evidence further suggests that sympatric distributions were typical for higher primates as early as 14 million B.P. or beyond. Whatever phylogenetic lineages are recognized, rapid hominid radiation during Pleistocene times took place within similar overall environments. There was, then, progressive speciation or at least character displacement, whether behavioral, ecological or morphological, within a sympatric range. This implies that Plio-Pleistocene hominid evolution took place mainly in relation to multiple, interfingering ecological opportunities. Presumably the cultural innovations, such as those reflected in early stone tool-making and animal butchery at sites such as Olduvai and Koobi Fora, would have laid the foundations for unprecedented species success of the most immediate human ancestor, greatly accelerating character displacement (Butzer, 1977).

Table 1. Depositional, topographic and regional interpretations for the Transvaal Australopithecine Sediments: the Sterkfontein Valley.

Swartkrans III (mid or late Pleistocene fauna [SKb]). Moderately long accumulation of clayey surface soils by surging runoff, with interruptions marked by minor flowstones. Implies deep soils and periodic heavy rains, with relatively effective ground vegetation outside. External conditions similar to those of today, possibly a little moister.

Swartkrans II (Acheulian artifacts and Homo; mid-Pleistocene fauna [SKb, early component], including butchery remains). Protracted accumulation of clayey surface soils by surging runoff or mass movements, with repeated interruptions marked by minor flowstones or corrosion. Implies deep soils and periodic heavy rains of considerable duration, with relatively effective ground cover. Moister than today.

Kromdraai KB (robust australopithecines, possible hominine butchery indicated; mid-Pleistocene fauna). Identical sedimentary record to Swartkrans II, except for absence of mass movements; possible overlap with Swartkrans II.

Sterkfontein 5 (Oldowan artifacts, possible hominines; mid-Pleistocene fauna [SE]). Moderately long accumulation of stoney surface soils by surging runoff in two phases, interrupted by minor flowstone. Implies shallow soils and high intensity rains, with incomplete ground cover. Conditions probably similar to today.

Kromdraai KA (early Pleistocene fauna). Relatively brief accumulation under conditions similar to those applying to Sterkfontein 5; possible partial overlap with Sterkfontein 5.

Swarkrans IC (robust australopithecine; early Pleistocene fauna in IB and IC [SKa]). Relatively brief accumulation of clayey surface soils by surging runoff and limited mass movements. Implies deep soils and periodic heavy rains of considerable duration, with relatively effective ground cover. Moister than today.

Swartkrans IB2 (robust australopithecines and Oldowan artifacts). Protracted accumulation of sandy surface soils by surging runoff, repeatedly interrupted by breaks with minor flowstone development. Implies thin soils and periodic high-intensity rains, with incomplete ground cover. Conditions similar to today.

Swartkrans IB1 (robust australopithecines, possible *Homo*, and Oldowan artifacts). Rapid accumulation of collapse breccia with clayey matrix soil. Implies deep soils and probably with a relatively effective ground cover. Similar to or moister than today.

Sterkfontein 4B-4C (gracile australopithecines; terminal Pliocene fauna [STS]). Moderately long accumulation of stoney surface soils by surging runoff, in two phases, interrupted by some erosion. Implies shallow soils and high intensity rains, with incomplete ground cover. Conditions probably similar to today.

Sterkfontein 4A (some fauna). Relatively rapid collapse of block breccia, followed by longer infilling of voids with clayey soil derivatives, interrupted by minor flowstone developments. Implies deep soils and periodic heavy rains, probably with relatively effective ground cover. Possibly moister than today.

Table 2. Interpretations for Australopithecine Sediments: Makapansgat

---

Makapansgat 5D, 5E. Protracted accumulation of clayey surface soils by surging runoff into open cave, with interruptions marked by flowstone laminae. Implies deep soil and periodic heavy rains, with relatively effective ground cover. Moister than today. Possibly overlaps with Swartkrans IB1.

Makapansgat 5A,5B,5C. Protracted accumulation of sandy surface soils by surging runoff into open cave. Rapid decline of clay level indicates destruction of soil mantle, with periodic, high-intensity rains and incomplete ground cover. Conditions apparently similar to today. Probably coeval with Sterkfontein 4B,4C.

Makapansgat 4. (Gracile australopithecine; termial Pliocene, cercopithecoid and rodent faunas). Protracted accumulation of clayey surface soils into small cave opening, in combination with internal collapse rubble. Implies deep soil and periodic heavy rains, with relatively effective ground cover. Moister than today, but possibly transitional. Probably overlaps with Sterkfontein 4A.

Makapansgat 3 (Gracile australopithecines; late Pliocene, bovid fauna; paleomagnetic interval ca. 2.9-3.2 million years). Repeated accumulation of fossiliferous marl lenses, interbedded with silts and flowstones, in shallow interlinked underground ponds; eventually cavern was sealed and extensive flowstones developed. Abundant subterranean waters linked to hydrology of a humid Karst river. Climate substantially moister than today. May antedate Sterkfontein 4A.

Makapansgat 2 (some fossiliferous units, not differentiated from unit 3; paleomagentic interval ca. 3.1-3.35 million years). Accumulation of silt lobes in an underground cavern by periodic lateral ingression of river discharge. Preponderence of fine suspended sediment infers hydrology of humid karst landscape, with reduced seasonal contrasts. Climate substantially moister than today.

Makapansgat 1. Long-term accumulation of massive, continuous flowstone under several meters of subterranean water seeping through cavern. During later stages, inruption of suspended sediment via underground fissures. Climate substantially moister than today with reduced seasonality.

References

Brain, C.K. 1981. *The Hunters or the Hunted?* Chicago: University of Chicago Press.

Butzer, K.W. 1974. Paleoecology of the South African australopithecines: Taung revisited. *Current Anthropology*, 15: 366-382, 413-415, 421-426.

Butzer, K.W. 1976. Lithostratigraphy of the Swartkrans Formation. *South African Journal of Science*, 72: 136-141.

Butzer, K.W. 1977. Environment, culture and human evolution. *American Scientist*, 65: 572-584.

Butzer, K.W. 1978. Geo-ecological perspectives on early hominid evolution. *In*: *African Hominidae of the Plio-Pleistocene*, ed., C.J. Jolly. London, Duckworth, pp. 191-217.

Butzer, K.W. 1980. The Taung australopithecine: contextual evidence. *Palaeontologia africana*, 23: 59-60.

Butzer, K.W. 1982. *Archaeology as Human Ecology*. New York: Cambridge University Press.

Butzer, K.W. n.d. Late Cenozoic environments in southern Africa as reflected in key fossiliferous and archaeological sites. In: *Southern African Palaeoenvironments and Prehistory*, ed., R.G. Klein. Rotterdam, Balkema, in press.

Butzer, K.W., Stuckenrath, R. Bruzewicz, A.J. and Helgren, D.M. 1978. Late Cenozoic paleoclimates of the Gaap Escarpment, Kalahari margin, South Africa. *Quaternary Research*, 10: 310-339.

Partridge, T.C. 1978. Re-appraisal of the lithostratigraphy of the Sterkfontein hominid site. *Nature*, 275: 282-287.

Partridge, T.C. 1979. Re-appraisal of the lithostratigraphy of the Makapansgat Limeworks hominid site. *Nature*, 279: 484-488.

Vrba, E.S. 1976. The fossil Bovidae of Sterkfontein, Swartkrans and Kromdraai. *Memoirs Transvaal Museum*, 21: 1-166.

AUSTRALOPITHECINES: THE UNWANTED ANCESTORS

Milford H. Wolpoff
University of Michigan

From the very beginning, the claim of an australopithecine ancestry for humans was not generally accepted. While the people, the arguments, and even the specimens have changed in the decades following the Taung discovery, this claim still forms the focus of one of the most important debates in paleoanthropology, and indeed one with the widest public interest since few problems either engender or deserve the attention that surrounds the studies of our origin.

Ramond Dart's interpretation of the Taung child was extraordinarily insightful, and with few exceptions the features he observed or inferred, and the context he provided for their interpretation, have withstood the test of time. However, his statements as published in 1925 and amplified in 1926 were only acceptable to a few scientists (Broom 1925a, 1925b; Sollas 1925, 1926; Romer 1930). Most recognized scholars instead regarded this child as an aberrant ape (Duckworth 1925; Keith 1925, 1931; Woodward 1925), perhaps varying somewhat in the hominid direction but evidently not an ancestral form.

Insight into this reaction can be gained from several sources (Brace and Montagu 1965; Dart with Craig 1959). I believe that an important role was played by the Piltdown "fossil", in blocking the door of hominid ancestry (LeGros Clark 1967). When it was reconstructed, the Piltdown skull could best be described as combining the braincase of a modern human with the jaws of an ape. The skull was large, with a high forehead and no browridge, but the jaw was chinless and did not seem different from a jaw of a living ape but for the fact that the canine was flattened from "wear." The find was accepted as being quite ancient because of its apparent association with extinct fauna. The implications of Piltdown were that at a very early time evolution produced the human brain, while the jaws and teeth "lagged behind," not developing a human form until much later. Theories of human origins had to account for an early evolution of the distinctive human mental development, and they were often along the lines of the idea that human evolution began when a particularly intelligent ape decided that a better life was to be had living on the ground (cf., Hooton, 1947; Keith, 1931; and others). With Piltdown accepted as a true ancient fossil hominid, it is little wonder that the discovery of the Taung australopithecine, and its interpretation as a human ancestor by Dart, was met with so much skepticism in 1925. According to Dart (1925), the australopithecines combined the jaws of a man with a braincase little modified from that of an ape. This is just the opposite of Piltdown, and logically both could not be human ancestors.

During the period when Piltdown was accepted as a true fossil, the australopithecines were almost universally regarded as being aberrant apes, or at best very early offshoots of the hominid line (Senyurek 1941; von Koenigswald 1942). Dart and Broom, the South African scholars, were alone in maintaining an ancestral position for these forms.

An interesting contrast was found in Weidenreich, who wrote extensively on the phyletic position of the australopitehcines during this period (1937, 1943, 1948) reaching the same conclusions as most other workers, although for very different reasons. Weidenreich stood virtually alone in not being swayed by the implications of Piltdown, since he did not regard it as a valid fossil. Yet, he considered most of the hominid-like features of the australopithecines as primitive retentions (1943), concluding that the group was a third major branch of the hominoids (along with hominids and pongids), with roughly equal relations to the other two:

> They must, therefore, be placed near the point where hominids and anthropoids proper diverged but, apparently, nearer to the gorilla branch than to any other of the three types (p. 272).

Weidenreich considered the hominid dental morphology as the ancestral condition, and the anthropoid morphology as derived. Therefore, he argued that to the extent that the australopithecine dentitions resembled hominids, this relation represented a primitive retention and was not useful in showing closeness of relationship. Similarly, other hominid-like features of the group (mandibular massiveness, cranial thickness, the deep mandibular fossa with an articular tubercle) represented the primitive condition (1948). In contrast, the fragmentary postcranial remains reported at the time (femora from Sterkfontein, a humerus, ulna, and talus from

Kromdraai) were so human-like that he regarded their association with the australopithecines as unlikely.

Thus, by reversing the very criteria for what was considered evidence of shared ancestry Weidenreich inferred an early divergence from the hominid-like features of the australopithecines, and a closer relation to the anthropoids from the shared simian features. In this regard, he drew a conclusion very similar to other non-South African workers of the time, but for entirely different reasons.

When Piltdown was revealed to be a hoax (Weiner 1955), the idea of an australopithecine ancestry for later hominids gained wider acceptance. In the early descriptive monographs on the accumulating South African finds of the 1930s and 1940s the South African workers strongly emphasized (or perhaps overemphasized) the human-like features of the fossils (Broom and Schepers, 1946; Broom, Robinson, and Schepers, 1950; Broom and Robinson, 1952). It would be fair to say that most scientists from this time until the early 1960s came to accept these arguments, and consequently the idea of an australopithecine stage to human evolution, with some, most, or all specimens on the direct line of human ascent.

One exception was Zuckerman, who entered into a series of colloquies; first with Le Gros Clark (1950a-e, 1951) questioning the hominid character of the australopithecine dentition (especially deciduous, see Ashton and Zuckerman 1950, 1951, 1952; Zuckerman 1950a-b, 1951a-c), and later with Robinson (1954a-b, 1958) questioning whether the occiputs then known supported the contention of erect posture as interpreted from a low nuchal line (Zuckerman 1954a-b; Ashton and Zuckerman 1956). The discovery of more complete crania seems to have terminated these arguments, but they were quickly rekindled with the publication of the virtually complete Sterkfontein pelvis (STS 14), and have continued virtually unabated to the present day (Zuckerman, 1966; Zuckerman et al., 1973). Zuckerman's concern was not a taxonomic one. Reflecting the general changes in the understanding of the australopithecines, he accepted the contention that the australopithecines were hominids, while rejecting the notion that these hominids were ancestral to living humans. This changed focus has continued to characterize the australopithecine debate as it developed through the second half of the century.

Von Koenigswald was (or more accutately, remained) another major exception during this period. While accepting the australopithecines as hominids once (1942), he consistently maintained that certain features of the australopithecine dentition, especially molarization of the dm_1, were too specialized for these forms to be hominid ancestors (see, for instance, von Koenigswald, 1967). Moreover, he claimed that these hominids were too late in time to be potential ancestors of *Homo erectus*.

While in a minority, these authors were not alone in arguing against any form of australopithecine ancestry for later hominids. For instance, Boule and Vallois (1957) regarded them as a less evolved branch of the primate stock. Chinese scholars have traditionally (Woo, 1962) and recently (Wang, Fang, and Zheng, 1981) regarded *Gigantopithecus* as the Asian version of *Australopithecus* and the true ancestor of later hominids. Thus, in spite of the growing appreciation for an australopithecine ancestry of later hominids, there has remained a continued tradition of scholars rejecting this interpretation.

Within the last 10 years, two major new arguments have been advanced to deny the concept of australopithecine ancestry for later hominids. The first of these reflects a position developed by Oxnard (1973, 1975, 1979). This position is based on two contentions: (1) the functional anatomy of the australopithecines involves many aspects that are neither human-like nor ape-like nor in between these two, and (2) co-existing with the australopithecines were true hominid ancestors.

The first part of this argument is based largely on Oxnard's original multivariate work, purporting to ascertain function from various statistical analyses either clustering or discriminating on the basis of metrics that are said to describe functional systems. Oxnard focused mainly on locomotor functions, although the same argument could be developed for the australopithecine dentition which in some respects (i.e., posterior tooth size) has also been claimed to be neither human-like nor ape-like nor between these two (Wolpoff 1973). Oxnard concludes that because the locomotor function of the australopithecines appears to be unique, their potential ancestry to later hominids must be suspect.

This approach, in fact, is only one example of a growing tendency to equate morphometrics with function.

Most workers seem to agree that the habitual stance of the australopithecines was bipedal. The impressive amount of evidence marshalled to support this point has been discussed in detail by workers such as Dart (1949, 1958), Lovejoy (1974, 1975), and others. Yet, within this context, Oxnard joins a number of authors in claiming that the australopithecines were characterized by gaits which differ from living humans, utilizing multivariate techniques in what I believe is a distortion of what Le Gros Clark meant by analysis of the "total morphological pattern." It is not by accident, in my view, that with the exception of work by Robinson (1972), every single study purporting to demonstrate gait differences in the early hominids has utilized multivariate procedures as the basis for what is then termed "functional analysis." The "functional" part of these analyses, however, is yet to be forthcoming, as is evidenced by the fact that none have indicated exactly _what_ the difference in gait involves.

Should this be surprising? I suspect not, in view of the total dependence of any multivariate approach on the assumptions which underlie the procedure (Kowalski, 1972), and the unfortunate extent to which the abstract results must be interpreted within a far removed biological context in the complete absence of any _a priori_ criteria with which one might ascertain the accuracy of this interpretation (Gnanadesikan, 1970).

There is no reason to believe that a multivariate analysis will necessarily reveal function (Wolpoff, 1976). Moreover, if Kowalski's (1972) stricture (that the value of such procedures can only be judged by their results) is correct, the work that has been done on fossil hominid material to date is at best discouraging. For instance, Day and Wood (1968) demonstrated to their satisfaction that the Olduvai Hominid 8 talus belonged to a biped incapable of the striding gait of living people, by use of a discriminant function. Oxnard (1973) was critical of the procedure, but later used a canonical analysis to argue that the OH 8 talus most closely resembles the fossil _Proconsul_ talus, and of the living pongids the closest resemblance is to _Pongo_ (1974). Since the resemblance is to the Orang, the "functional" interpretation is an adaptation to acrobatic arboreal climbing. In this case there is some basis for judging the accuracy of these biological interpretations, since the talus in question is one of 12 bones forming the better part of the OH 8 foot (Day, 1976). When subject to analysis, the foot reveals all those features commensurate with a striding gait (Archibald, Lovejoy, and Heiple, 1972). The verdict for the multivariate procedures is not favorable. A talus is not a foot.

Not surprisingly, the hypothesis testing approach has resulted in a completely different interpretation of australopithecine locomotion. In a series of sophisticated morphological analyses, Lovejoy and his co-workers have shown that australopithecine pelvic morphology can be adequately explained under the hypothesis of a completely human gait repertoire combined with a smaller-than-human pelvic inlet (1974, 1975; Lovejoy, Heiple, and Burstein, 1973). Moreover, the combinations of widely flared ilia and narrower inlet provide an explanation for the main distinctive feature of australopithecine femora, the relatively long neck joining the shaft to the more medially placed head. To date this functional interpretation is yet to be refuted on morphological grounds. Demonstrations that the various australopithecine limbs do not sort with living humans on the basis of multivariate morphometric analyses are not refutations of this hypothesis or any other, since these demonstrations are yet to be interpreted functionally.

Apart from these comments, there is in my view a more basic fallacy in Oxnard's approach. Why should one expect the australopithecine locomotor pattern to conform to that of living humans, or living pongids, or for that matter be in some sense between them? Living pongids did not evolve into living humans (or vice versa). Why should the morphological features of human ancestors be halfway between living humans and middle Miocene apes? Why, in fact, should some features be anywhere between them? Fossils are examples of once living successful populations, and their adaptive response to the environment may be quite different from either their ancestors or their descendents. There is no doubt that, as Oxnard states, "the fossils . . . exhibit anatomical arrangements that render them unique." To suggest otherwise would be to deny evolutionary change Whether their unique aspects render them unacceptable as human ancestors depends on the prior and subsequent course of evolution. In the case of the australopithecines, in some respects they are ape-like, in others very human-like, and in yet others they are unlike apes, humans, or anything "in-between."

Oxnard's second contention, that co-existing with the australopithecines were true hominids, is a familiar one in that it is a specialized version of the "australopithecines were too late" argument that has appeared throughout the entire history of the ancestry issue. In this case, Oxnard has relied on contemporaneity based on the now discredited 2.6 million year date for the KBS tuff (Cerling *et al.*, 1979; Drake *et al.*, 1980) and the guesstimation that was made for the Kanapoi humerus (Patterson and Howells 1967). With the more recent dating of the KBS tuff, and the accumulating evidence that most if not all of the australopithecine-bearing sites from South Africa predate it (White and Harris, 1977), it appears possible that all and probably that most of the australopithecines predate the appearance of remains attributed to "Homo" in East Africa (Walker and Leakey, 1978).

The second and most recent of the new arguments against an australopithecine ancestry for humans has two similar variants. The first of these, probably best presented by R. E. Leakey (1981; Leakey and Lewin, 1977), is a resurrection of the widespread opinion common to the decades following the Taung discovery. Leakey asserts that no australopithecine contributed to human ancestry, contending that during the entire timespan of australopithecine existence, a rarer but more advanced hominid was present. While not disproved, and perhaps not disprovable, this variant has also not been found convincing to many. The problem is that the more advanced hominid is widely believed to be an australopithecine itself (for instance Walker, 1976; Wolpoff, 1980) with observable variation in the consequent australopithecine range interpreted as a result of sexual dimorphism and temporal and geographic distinctions.

The second variant is a modification in that the very earliest australopithecines are retained as hominid ancestors, while all others are removed. This argument was presented as part of the initial discussion concerning the validity of the newly named species *Australopithecus afarensis* (Johanson and White, 1979), and in effect dismissed virtually all known australopithecines from human ancestry except for those described in the new taxon by these authors. Widely heralded as the new theory of hominid evolution, their approach (White, Johanson, and Kimbel 1981) actually represents a fleshing out of von Koenigswald's ideas (although interestingly if von Koenigswald's interpretation of deciduous molar evolution were applied to *Australopithecus afarensis*, these hominids would probably also be relegated to the extinct side branch of australopithecine evolution).

This scheme admits the likelihood that the South African sequence represents a single evolving lineage (Wolpoff 1973b) from Sterkfontein Member 4 and Makapan to Swartkrans Member 1 and Kromdraai. It is argued that from the very beginning, the South African hominids show an approach to the Australopithecus boisei condition, and consequently are too derived to be ancestral to Homo erectus. However, at the same time the authors found themselves defending the species distinction they proposed for their sample. Tobias (1980) for instance, believes the separation from Australopithecus africanus to be invalid. In response, it has been contended that in comparisons of features that vary between these samples, Australopithecus afarensis is almost invariably the more primitive of the two.

One would think that these two arguments are complimentary, but this does not always prove to be the case. Obviously, in areas where Australopithecus afarensis is more primitive, Australopithecus africanus must be more derived, but in a good number of these the resemblance is to later hominids recognized to be in the genus Homo by the authors (Homo habilis, Homo erectus), and these comparisons are not mentioned in the discussion purporting to show that Australopithecus africanus is derived in the direction of Australopithecus boisei. This is not a trivial point, since the longer the list of shared derived features, the more likely it is that the latest common ancestor of the two Pleistocene hominid lineages postdates Australopithecus africanus.

In the postcranial skeleton, it is clear that all of the australopithecines are erect, striding bipeds. However, certain morphological aspects of the skeleton of the later australopithecine species clearly vary in the direction of Homo. This pattern has been published for the humerus (compare Senut 1981; Broom and Schepers, 1946) and probably also characterizes the radius and ulna. Although these comparisons did not appear in the formal diagnosis of Australopithecus afarensis, they probably should have.

In the dentition, derived and specifically Homo-like features in the later australopithecines include the consistent appearance of a bicuspid P_3, the loss of the triangular M_3 occlusal form, reduction of deciduous canine crown height and pointedness, reduction in permanent incisor and canine size, reduction in the mammalon incisors, the appearance of symmetric labial crown profiles in the upper canines with characteristically apical wear, the appearance of symmetry in the occlusal form of the P^3, and reduction in buccal cusp height in the permanent upper molars.

In the mandibles, Australopithecus africanus features derived in the direction of Homo include a more vertical and straighter external symphysis, anterior and superior migration of the inferior transverse torus (a structure retained in many Homo erectus mandibles), a more posterior angulation for the P_3 roots, and the loss of an anterosuperior direction to the mental foramen opening.

In the facial skeleton, Australopithecus afarensis is extraordinarily primitive, and clearly adapted for a diet with elements that required powerful mastication. In the premaxillary and anterior maxillary region, the later australopithecines show Homo-like variation in reductions associated with the diminuation of large canines and large central incisors with long curved roots. However, many other contrasts reported between the earlier and later australopithecine taxa loose significance in the light of what appears to be normal variation within the genus Homo. Differences reported in the position of the anterior roots of the zygomatic processes, the insertion of the volmer relative to the anterior nasal spine (and indeed the presence of this spine as a distinct structure), and the relation between the floor of the nose and the external face of the premaxilla all find their counterparts in modern populational variation, and in many cases in the small Homo erectus facial sample as well. In general, apart from the premaxillary region australopithecine faces show much more similarity than contrast when considered over time. The same can be said for the supraorbital region of the frontal as represented in the early sample by a fragment of AL 288-1 and a newly discovered specimen reported in the press by T. D. White; the australopithecine condition characterizes both Australopithecus afarensis and Australopithecus africanus, and differs from the corresponding region in later Homo habilis (ER 1813, OH 16).

It may well be in the cranial vaults that Australopithecus africanus shows the greatest number of features derived in the direction of Homo. Many of these are associated with a reduction of the nuchal musculature (following from a decrease in horizontally oriented anterior loading) and an average forward shift in the foramen magnum position. Thus, in *Australopithecus africanus* parallel temporal lines do not extend onto the occiput. The related question of a compound temporal/nuchal crest in *Australopithecus africanus* is probably irrelevant since the structure appears in later hominids (ER 1805) and occasionally even in modern humans (Robinson 1958). I have reported (Wolpoff 1974, p. 400) that certain *Australopithecus africanus* males such as STS 71 approach this condition with "an apparent, although not raised, compound crest area" (contra claims by White, Johanson, and Kimbel 1981, p. 456). The difference in this region between the earlier and later australopithecines results from a combination of smaller nuchal muscles and a markedly expanded occipital plane in *Australopithecus africanus*, and in these respects it more closely resembles *Homo*.

The inferior temporal surface, including the mandibular fossa, is another area in which *Australopithecus africanus* is clearly more derived, approximating or foreshadowing the condition in *Homo* species. The roof of the fossa is high, and the articular surface is angled, with a clear posteroinferiorally oriented face. Other differences lie in the morphology of the tympanics and the mastoid process (especially its posterior face).

In sum, the arguments for species distinction turn out to be a two edged sword. The very case that establishes *Australopithecus afarensis* as a distinctly more primitive australopithecine species, also reveals *Australopithecus africanus* to possess a suite of derived features shared with early members of the lineage leading to *Homo sapiens*. That some of these are also shared with the other Pleistocene hominid lineage (a few of these uniquely so) is best explained by the contention that *Australopithecus africanus* is the latest common ancestor.

For more than a half-century, various scholars have attempted to dismiss the australopithecines from hominid ancestry. Indeed, such attempts characterize both the very earliest and the very latest discussions of the problem. There is every reason to believe they will continue. I doubt there is a single all encompassing reason for this. Elements of personal motivation, differences in what features are judged to be primitive, varying interpretations of the evolutionary process (especially concerning expectations of how much time is required for changes to occur), the expectation that evolution should precede in a straight line as contrasted with the expectation that the process is mossaic and each species is unique, differing experiences with normal populational variation, disagreements about which analytical techniques are appropriate, and the very nature of paleontological evidence which is so often fragmentary, less than adequately dated, and virtually never representative of actual biological populations, all play a role in the inability of paleoanthropology to settle this issue. And perhaps this is for the better. Scientific progress is a result of progressive refutations. What could better serve to insure the continued critical re-examination of the australopithecine ancestry hypothesis than the persistent publications of workers attempting to refute it? The longer such attemps fall short of success, the more confident we may feel that the bulk of the hominid fossil record represents the remains of our own ancestors, and thus tells the story of our evolution.

Acknowledgments

For their permission to study the specimens in their care, I am very grateful for the help and encouragement provided by C.K. Brain, L. Vrba, P.V. Tobias, A. Hughes, M. Leakey, R.E. Leakey, Y. Coppens, R. Protsch, D.C. Johanson and W. Kimbel. This work was supported by NSF Grant BNS 76-82729.

References

Archibald, J.D., C.O. Lovejoy, and K.G. Heiple. 1972. Implications of relative robusticity in the Olduvai metatarsus. American Journal of Physical Anthropology 37: 93-95.

Ashton, E.H., and S. Zuckerman. 1950. Some quantitative dental characters of fossil anthropoids. Philosophical Transactions of the Royal Society, Series B, 234: 485-520.

Ashton, E.H., and S. Zuckerman. 1951. Some dimensions of the milk teeth of man and the living great apes. Man 51: 23-26.

Ashton, E.H., and S. Zuckerman. 1952. Overall dental dimensions of hominoids. Nature 169: 571-572.

Ashton, E.H., and S. Zuckerman. 1956. Cranial crests in the anthropoidea. Proceedings of the Zoological Society of London 126: 581-625.

Brace, C.L., and M.F.A. Montagu. 1965. Man's Evolution. MacMillan, New York.

Broom, R. 1925a. Some notes on the Taungs skull. Nature 115: 569-571.

Broom, R. 1925b On the newly discovered South African man-ape. Natural History 34: 409-418.

Broom, R., and J.T. Robinson. 1952. Swartkrans Ape-man. Transvaal Museum Memoir 6, Pretoria.

Broom, R., J.T. Robinson, and G.W.H. Schepers. 1950. Sterkfontein Ape-Man Plesianthropus. Transvaal Museum Memoir 4, Pretoria.

Broom, R., and G.W.H. Schepers. 1946. The South African Fossil Ape-Men, the Australopithecinae. Transvaal Museum Memoir 2, Pretoria.

Cerling, T.E., F.H. Brown, B.W. Cerling, G.H. Curtis, and R.E. Drake. 1979. Preliminary correlations between the Koobi Fora and Shungura formations, East Africa. Nature 279: 118-121.

Dart, R.A. 1925. Australopithecus africanus: the man-ape of South Africa. Nature 115: 195-9.

Dart, R.A. 1926. Taungs and its significance. Natural History 26: 315-327.

Dart, R.A. 1949. Innominate fragments of Australopithecus prometheus. American Journal of Physical Anthropology 7: 301-333.

Dart, R.A. 1958. A further adolescent ilium from Makapansgat. American Journal of Physical Anthropology 16: 473-9.

Dart, R.A., with D. Craig. 1959. Adventures with the Missing Link. Viking, New York.

Day, M.H. 1976. Hominid Postcranial material from Bed I, Olduvai Gorge. In: Human Origins: Louis Leakey and the East African Evidence, eds. G.L. Isaac and E.R. McCown W.A. Benjamin, Menlo Park. pp. 363-374.

Day, M.H. and B.A. Wood. 1968. Functional affinities of the Olduvai Hominid 8 talus. Man 3: 440-445.

Drake, R.E., G.H. Curtis, T.E., Cerling, B.W. Cerling and J. Hampel. 1980. KBS Tuff dating and geochronology of tuffaceous sediments in the Koobi Fora and Shungura formations, East Africa. Nature 283: 368-371.

Duckworth, W.L.H. 1925. The fossil anthropoid ape from Taungs. Nature 115: 236.

Gnanadesikan, R. 1970. S.N. Roy's interests in and contributions to the analysis and design of certain quantitative multiple response experiments. In: Essays in Probability and Statistics, eds. R.C. Rose, I.M. Chakravarti, P.C. Mahalanobis, C.R. Rao, and K.J. Smith. University of North Carolina Press, Chapel Hill. pp. 293-310.

Hooton, E.A. 1947. Up From the Ape, revised edition. MacMillan, New York.

Johanson, D.C., and T.D. White. 1979. A systematic assessment of early African hominids. Science 202: 321-330.

Keith, A. 1925. The fossil anthropoid ape from Taungs. Nature 115: 234-235.

Keith, A. 1931. New Discoveries Relating to the Antiquity of Man. Williams and Norgate, London.

von Koenigswald, G.H.R. 1942. The South-African Man-Apes and Pithecanthropus. Carnegie Institute of Washington Publication 530: 205-222.

von Koenigswald, G.H.R. 1967. Evolutionary trends in the deciduous molars of the Hominidea. Journal of Dental Research 46: 779-786.

Kowalski, C.J. 1972. A commentary on the use of multivariate statistical methods in anthropometric research. American Journal of Physical Anthropology 36: 119-132.

Leakey, R.E. 1981. The Making of Mankind. Dutton, New York.

Leakey, R.E., and R. Lewin. 1977. Origins. Dutton, New York.

LeGros, Clark, W.E. 1950a. South African fossil hominids. Nature 165: 791.

LeGros, Clark, W.E. 1950b. South African fossil hominids. Nature 165: 893.

LeGros, Clark, W.E. 1950c. New discoveries of the australopithecines. Nature 166: 758-760.

LeGros, Clark, W.E. 1950d. South African fossil hominids. Nature 166: 791-792.

LeGros, Clark, W.E. 1950e. Hominid characters of the australopithecine dentition. Journal of the Royal Anthropological Institute of Great Britain and Ireland 80: 37-54.

LeGros, Clark, W.E. 1951. Comments on the dentition of the fossil Australopithecinae. Man 51: 18-20, 32.

LeGros, Clark, W.E. 1967. Man-apes or Ape-men? The Story of Discoveries in Africa. Holt, Rinehart, and Winston, New York.

Lovejoy, C.O. 1974. The gait of australopithecines. Yearbook of Physical Anthropology 17: 147-161.

Lovejoy, C.O. 1975. Biomechanical perspectives on the lower limb of early hominids. In: Primate Functional Morphology and Evolution, ed. R. Tuttle. Mouton, The Hague. pp. 291-326.

Lovejoy, C.O., K.G. Heiple, and A.H. Burstein. 1973. The gait of Australopithecus. American Journal of Physical Anthropology 38: 757-780.

Oxnard, C.E. 1973. Functional inferences from morphometrics: problems posed by uniqueness and diversity among the primates. Systematic Zoology 22: 409-424.

Oxnard, C.E. 1975. The place of the australopithecines in human evolution: grounds for doubt? Nature 258: 389-396.

Oxnard, C.E. 1979. Relationship of Australopithecus and Homo: another view. Journal of Human Evolution 8: 427-432.

Patterson, B., and W.W. Howells. 1967. Hominid humeral fragment from early Pleistocene of Northwestern Kenya. Science 156: 64-66.

Robinson, J.T. 1954a. The australopithecine occiput. Nature 174: 262-263.

Robinson, J.T. 1954b. Nuchal crests in australopithecines. Nature 174: 1197-1198.

Robinson, J.T. 1958. Cranial cresting patterns and their significance in the Hominoidea. American Journal of Physical Anthropology 16: 397-428.

Robinson, J.T. 1972. Early Hominid Posture and Locomotion. University of Chicago, Chicago.

Romer, A.S. 1930. Australopithecus not a chimpanzee. Science 71: 482-483.

Senut, B. 1981. Humeral outlines in some hominoid primates and in Pliopleistocene hominids. American Journal of Physical Anthropology 56: 275-284.

Senyurek, M.S. 1941. The dentition of Pleisanthropus and Paranthropus. Annals of the Transvaal Museum, 20: 293-302.

Sollas, W.J. 1925. The Taungs skull. Nature 115: 908-909.

Sollas, W.J. 1926. A sagittal section of the skull of Australopithecus africanus. Quarterly Journal of the Geological Society 82 (pt. 1): 1-11.

Tobias, P.V. 1980. "Australopithecus afarensis" and A. africanus: critique and an alternative hypothesis. Palaentologia Africana 23: 1-17.

Walker, A. 1976. Remains Attributable to Australopithecus in the East Rudolf Succession. In: Earliest Man and Environments in the Lake Rudolf Basin, eds. Y. Coppens, F.C. Howell, G.L. Issac, and R.E.F. Leakey. University of Chicago Press, Chicago pp. 484-489.

Walker, A., and R.E.F. Leakey. 1978. The hominids of East Turkana. Scientific American 239 (2): 54-66.

Wang Linghong, Fang Kaitai, and Zheng Yuying. 1981. An application of Bayes discrimate analysis in determining the systematic position of Gigantopithecus, Vertebrata PalAsiathica 19: 269-275.

Weidenreich, F. 1937. The dentition of Sinanthropus pekinensis: a Comparative odontography of the hominids. Palaeontologica Sinica, New Series D, 1 (Whole Series 101): 1-180.

Weidenreich, F. 1943. The Skull of Sinanthropus pekinensis: A comparative study of a primitive hominid skull. Palaeontologia Sinica, n.s. D, No. 10 (Whole series No. 127).

Weidenreich, F. 1948. About the morphological character of the australopithecine skull. Robert Broom Commemorative Volume, ed. A. du Troit. Royal Society of South Africa, Capetown. pp. 153-158.

Weiner, J.S. 1955. The Piltdown Forgery. Oxford, London.

White, T.D., and J.M. Harris. 1977. Suid evolution and correlation of African hominid localities. Science 198: 13-31.

White, T.D., D.C. Johanson, and W.H. Kimbel. 1981. Australopithecus africanus: its phyletic position reconsidered. South African Journal of Science 77: 445-470.

Wolpoff, M.H. 1973a. Posterior tooth size, body size and diet in South African gracile australopithecines, American Journal of Physical Anthropology 39: 375-394.

Wolpoff, M.H. 1973b. The evidence for two australopithecine lineages in South Africa. Yearbook of Physical Anthropology 17: 113-139.

Wolpoff, M.H. 1974. Sagittal cresting in the South African australopithecines. American Journal of Physical Anthropology 40: 397-408.

Wolpoff, M.H. 1976. Multivariate discrimination, tooth measurements, and early hominid taxonomy. Journal of Human Evolution 5: 339-344.

Wolpoff, M.H. 1980. Paleoanthropology. Knopf, New York.

Woo, Ju-Kang. 1962. The mandibles and dentition of Gigantopithecus. Palaeontologia Sinica, New Series D, Number 11.

Woodward, A.S. 1925. The fossil anthropoid ape from Taungs. Nature 115: 235.

Zuckerman, S., E.H. Ashton, R.N. Flinn, C.E. Oxnard, and T.F. Spence. 1973. Some locomotor features of the pelvic girdle of primates. Proceedings of the Zoological Society of London 33: 71-165.

Zuckerman, S. 1950a. South African fossil hominids. Nature 166: 158-159.

Zuckerman, S. 1950b. South African fossil hominids. Nature 166: 953-954.

Zuckerman, S. 1951a. Comments on the dentition of the fossil Australopithecinae. Man 51: 20

Zuckerman, S. 1951b. The dentition of the Australopithecinae. Man 51: 32.

Zuckerman, S. 1951c. An ape or the ape? Journal of the Royal Anthropological Institute 81: 57-65.

Zuckerman, S. 1954a. The australopithecine occiput. Nature 174: 263-264.

Zuckerman, S. 1954b. Nuchal crests in australopithecines. Nature 174: 1198.

Zuckerman, S. 1966. Myths and methods in anatomy. Journal of the Royal College of Surgeons of Edinburgh 11: 87-114.

THE KOHL-LARSEN EYASI AND GARUSI HOMINID FINDS IN TANZANIA AND THEIR RELATION TO HOMO ERECTUS

R.R.R. Protsch
J. W. Goethe University
Federal Republic of Germany

Although the Garusi Hominid presents a more questionable placement in literature, it is safe to say that traditionally the Eyasi and Garusi Hominids have been attributed to *Homo erectus*. Weinert who was the main proporter of their taxonomic placement quite probably worked more in accordance with preconception rather than the usual scientific procedure. It is often stated that his work on this fossil material was meticulously exact and extensively carried out. Even after a short examination of the original reconstruction as well as of additional pieces it should, however, become quite obvious to the trained morphologist that Weinert's reconstruction is questionable.

A new reconstruction of these finds has been attempted. In its present state a detailed description of new details is possible and provides new clues as to its taxonomic status (Protsch, 1977). It is the intention of the author to give a short history of these finds in a relatively brief form including a short description, their state of preservation, number and the importance of additional pieces and their assignment to specific individuals (i.e. Eyasi I, II, III, IV). Dates or possibility of dating, as well as suggestions as to taxonomic grouping will also be touched upon. A general comment will be followed by a specific section dealing with first the Eyasi Hominids and then the Garusi. Compared to other geographical areas of the world East Africa and specifically the area around and close to Olduvai Gorge has yielded only scarce material of those hominids which could be assigned to or within the *Homo erectus* group. At a time when only the hominid find of Olduvai Gorge known as Hominid 1 (OH 1) which was discovered in 1913 and a few somewhat more recent dating hominid finds (ca. up to 15,000 years B.P.) were known, two finds aroused considerable interest among palaeoanthropologists. These finds were those from Eyasi and Garusi which were discovered by the German Kohl-Larsen Expedition of East Africa during several excavation seasons spanning from 1934 to 1940 (Kohl-Larsen, 1943; Reck and Kohl-Larsen, 1936).

Today new finds from this area, some of older chronological order, are readily available to anthropologists either in the form of casts or original material. In the past, however, only few specialists had access to such originals and the production of casts for other specialists was generally neglected. It is under these circumstances not surprising that a few mistakes were made by these few anthropologists who studied these finds; mistakes which could only be corrected many years later. A few photographs could give only limited information about the morphology of these finds and the restudy of some originals was often hampered by their removal from institutions for safekeeping and later ultimate disappearance, as in the case of the Eyasi Hominids during World war II.

When in the 1960's the originals were recovered again new studies of their morphology in the early 1970's included a new reconstruction and the application of new methods to this osteological material, like chemico-physico-types of dating, serving to give us a more complete and correct picture of these fossil hominids. The fossil hominids were in the late 1930 and early 1940's, after only short and preliminary morphological investigations, too readily assigned to a group of fossil hominids which were at that time not yet known to exist in East Africa or for that matter not on the whole African Continent. Their study by H. Weinert and others were only seemingly extensive (Weinert, 1937; 1938; 1938a; 1938b; 1938c; 1940; 1952). Weinert's hasty assignment of these finds to the _Homo erectus_ or pre-_Homo erectus_ groups, even before their reconstruction, should leave some doubt as to the correctness of their proposed position in hominid phylogeny. The Eyasi Hominids consisted of so many fragmentary pieces (over 250) that only a very diligent and lengthy reconstruction could solve the question of their taxonomic and chronological position. In the case of the second hominid find from this East African area, namely Garusi, only a small part of the right maxilla with P^3 and P^4 in place as well as a separate molar (M^3) was found with the alveoli. The canines and incisor in the former were still well preserved though covered with matrix until 1973. Previous scholarly evaluations were dependent on only a few poorly made casts which gave a somewhat incorrect picture and hampered an exact morphological evaluation.

With the availability of new methods and retrieval of additional fossil bone material it was possible to reconstruct and evaluate both hominids more correctly. These included the use of more delicate cleaning equipment and absolute dating for precise chronological placement.

To be more explicit the following points make it now possible to assign both hominids positively into more specific fossil hominid groups:

1) New morphological reconstruction
2) Removal of matrix covering important parts of the fossils
3) Relative as well as absolute placement in time
4) Morphological comparison to fossils of the same locality and stratigraphy
5) Mineralogical evaluation of the matrix adhering to these fossils and its comparison to new hominid finds with the same morphology and of the same stratigraphy
6) Application of indirect A_1-dating by Potassium/Argon of the new Laetolil hominids to some of above fossil hominids (A_3-date).

The Eyasi Hominids

Contrary to accounts in most text books the Eyasi Hominids were not lost during WW II but stored on a farm for safekeeping. In the late 1960's they were relocated at Tubingen, West Germany. Relocated were all hominids presently known as Eyasi I, II, III, and IV. Eyasi II, an occipital fragment, was returned to Dar-es-Salaam, Tanzania, several years thereafter. All of the fragments, including some associated faunal material, were originally found during two excavations seasons from 1934 to 1940. The reason why a German expedition was still doing research in this area is that it was a former German colony (German East Africa). This meant that they had detailed knowledge of the region through a series of expeditions by the military, geologists, and a number of other scientists during the period from 1876 to 1914. L.S.B. Leakey, on the other hand had not yet started an extensive study of the area.

Lake Eyasi is situated 50 to 55 km south of Olduvai Gorge and the area of the fossil finds is located in a bay at the North-east site of the lake. The bay itself is erroneously mentioned in the Kohl-Larsen expedition report as the western bay. The excavation of the locality was initially to be only a small project. Far greater concentration was placed on the expedition's major goal which was the collection of ethnological material on a tribe called the Tindiga inhabiting an area around Lake Eyasi (then called Lake Njarasa by the Germans).

All excavated materials were surface to subsurface finds contained in a sandstone layer to a depth of about 35 cm. Hominids, fauna, and artefacts were found randomly mixed together in this layer which is located directly on the lake shore. Even though a more extensive excavation was planned because of the initial promising finds groundwater danger and the migration of surface waters, which frequently came within 30 to 110 m of the site, hindered a more intensive search for additional material. The major hominid find, Eyasi I, was discovered on the 29th of November 1935. It was assigned a relatively chronological age on the basis of fauna, artefacts and appearance - the bones being heavily permineralized.

This quite fragmentary material which was found scattered over an extensive area, numbered over 200 fragments. They were used in a preliminary reconstruction by H. Reck and L.S.B. Leakey in 1936. Even though Leakey assigned the specimen Eyasi I, according to its primitive characteristics, to a fairly early time period and possibly closely related to the Pithecanthropus group, he cautioned against using solely this criterion (Leakey 1936, 1946, 1948, 1952). He believed that associated finds had to be considered and a more detailed reconstruction had to be attempted first. In light of Leakey's cautious consideration of the facts it is especially surprising that Weinert (1937), after only a short examination, unhesitatingly gave the fossil the name _Africanthropus njarasensis_, a nomen nudem already assigned to the Florisbad Hominid. He positively aligned it with Pithecanthropus. In doing so Weinert not only disregarded Leakey's, Reck's and Kohl-Larsen's thorough investigations of geological deposits in which the fossils were found but at the same time tried to defend himself against criticism by F. Birkner (1937). Birkner saw Eyasi as a possible member of the Neanderthal in Africa. When Bauermeister (1940), Weinert's assistant, attempted a new reconstruction in 1939 he had changed little from the original one by Leakey and Reck. Only a few additions and alterations were made on the frontal and occipital. These parts were, however, some of the most decisive ones in assigning it to either _Homo erectus_ or to the African Neanderthals. To be more specific, very few of the fragments except three in the region of the parietals fit together exactly in Weinert's reconstruction. Most pieces are separated by large gaps giving the entire original reconstruction a somewhat disjointed and awkward appearance. The resultant morphological picture resembled Homo erectus in numerous aspects. Numerous small pieces found separately couldn't possibly be included in Weinert's reconstruction in that state. Weinert has pointed out (1938c) that the hominid remains are remnants of cannibalism. He based his assumption on the quite fragmentary nature of the bones. There is no evidence which would support such a theory concerning the Eyasi hominids. After microanalysis (F,U,N), these fragments could easily be assigned to specifically Eyasi Hominid I, II, III, or IV. This relative placement technique was not available to the original researchers of the 1930's and early 1940's (Protsch, 1976), the value of which in this case is quite critical. For example some pieces because they were found as far away as 430 meters from the original site of the 1934 excavation had then been immediately assigned to a hominid other than Eyasi I. All of these are smaller fragments with more Diploe than Lamina externa or Lamina interna and were thus of light weight. This indicates that they were probably transported by water and wind in a south-easterly direction and than redeposited there. With microanalysis it is then possible to assign the many separate pieces to individual hominids. Indeed, most of them belong specifically to Eyasi I, according to microanalysis. This enabled us to add additional parts to this hominid and allowed us to attempt a more complete new reconstruction.

Palaeoanthropus njarasensis - the second name given to the Eyasi Hominids - seems to belong to the "African Neanderthal" group, Homo sapiens rhodesiensis, rather than to Homo erectus. Ultimately, absolute dates assigned to the fossil hominid Eyasi I by amino-acid dating run by two different laboratories designate dates of 34,000 and 35,000 years B.P. (Aspartic acid) to these hominids (Protsch, 1975). These dates further support the assignment of the hominid to Homo sapiens rhodesiensis and allow a proper taxonomic evaluation on the basis of the new morphological reconstruction. To strengthen this argument it should also be pointed out that there is, to date, no palaeoanthropological evidence to support the existence of Homo erectus in this late time period.

Eyasi I and possibly the other Eyasi hominids as well can thus be added to the other members of Homo *sapiens rhodesiensis* which include Saldanha, dating to ca. 40,000 years B.P. (Protsch, 1975) and Broken Hill, possibly dating to ca. 110,000 years B.P. (Bada, Protsch, 1974). The Broken Hill date, also an amino-acid date, should be used with caution and viewed with scepticism. As in all cases this technique should be used in conjunction with other supportative absolute ones. Besides temperature and water amino-acid dating might also be influenced by associated metals, as is the case with Broken Hill.

The morphology of Eyasi I is not too different from Saldanha and Broken Hill but quite dissimilar to *Homo erectus*; if there is such a thing as "typical *Homo erectus* morphology." According to initial estimates the cranial capacity of Eyasi I is about 1220 to 1250 ccm, a measurement which is close to that of the Broken Hill individuals. The skull curvature seems also to align this hominid with Broken Hill, whereas the occipital does not differ substantially from the late Middle Pleistocene hominid of Steinheim (also Weinert, 1940). An alignment according to anthropological Normae shows that in Weinert's reconstruction neither Norma occipitalis, nor Norma frontalis, nor Norma parietalis, nor Norma verticalis can be correct. Is is impossible even to see a continuous Occipital torus or a Supra-orbital torus, as well as tight Post-orbital-constriction which is typical of *Homo erectus*. The skull is not as platycephalic and the Nuchal crest is not at all horizontal as is characteristic of *Homo erectus*. The maximum skull breadth does not seem to occur towards the lower part of the skull on the temporals but instead high on the parietals. The thickness of the cranial bones is normal and falls into that of anatomically modern man. The above observations are supportative for exclusion of the specimen from the *Homo erectus* group.

The internal morphology which was completely neglected by Weinert gives after removal of all plaster parts valuable additional information which could be used to refit pieces properly. In many cases the Lamina interna was well preserved, contrary to Weinert's assertion (Weinert et al. 1940) and could be used for proper fitting. Besides this over fifty additional pieces could now be utilized in the new reconstruction. This allowed a more proper and accurate fitting of sutures and pieces such as a part of the left maxilla with a canine and premolar (P^3) belonging to Eyasi I.

Many pieces of the former Eyasi II and Eyasi III are faunal material; other pieces are very fragmentary hominid material with only a few pieces preserved well enough to recognize morphological features justifying an inclusion in Homo sapiens rhodesiensis (Protsch, 1977).

On the basis of morphology the Eyasi individuals should be separated from other Homo erectus individuals of East Africa. The Eyasi hominids are neither similar to Olduvai Hominid 9 (OH 9) nor, judging from publication photographs, to the new Ndutu specimen assigned to Homo erectus, a claim made by Clark (1976). It is advisable that more detailed analysis and comparison should be attempted in the near future.

The Eyasi hominids do not seem to be the only hominids excavated since 1940 around Lake Eyasi. In 1967 a short excavation season was conducted by J. Ikeda of Kyoto University, Japan, at Bangani, Lake Eyasi. Even though no detailed morphological descriptions have been published yet. Ikeda believes that the two Bangani mandibles excavated are contemporary with the Eyasi I hominid (pers. comm. Ikeda, 1975).

On geological, palaeontological and cultural grounds as well as relative and absolute chronological comparisons it seems impossible that the Eyasi hominids could date anywhere between 600,000 and 500,000 years B.P. as seems to be the case with the Ndutu individuals (Mturi, 1976).

To conclude, the author's findings are supportative of the original interpretation of Leakey, Reck and Kohl-Larsen based on geological, cultural and faunal evidence for an Upper Pleistocene age of around 35,000 years B.P. (Reeve, 1946).

The Garusi (Ngarusi) Hominid

This hominid was found in February 1939 on the northwest side of Lake Eyasi inside a gorge of the Garusi river which flows into Lake Eyasi. This location is closer to Olduvai Gorge than the one where the Eyasi hominids were found. The excavation was rather short since war was imminent between England and Germany and the German expedition was under constant fear of interruption by British authorities. Contrary to reports in literature more material than the famous right maxillary fragment with the P^3 and P^4, commonly known as Meganthropus africanus, was found. Under these conditions on February 20, skull fragments were found. Among these was a larger occipital fragment which was subsequently lost in Germany. The somewhat different state of fossilization seemed to indicate that the latter piece could not be aligned with the maxillary fragment found earlier.

On the 24th of February 1939, a separate Molar (M^3) and another occipital fragment were found. Since the second occipital fragment displayed again another unique state of fossilization it too was discarded and only the molar was kept. These finds came from the same tuffaceous deposit now known as the Laetolil Beds, situated close to the Deturi (Laetolil) River. The separate M^3 was first thought to be part of the same individuals to which the maxillary fragment belonged (Abel, 1940). Later investigators, however, excluded the molar from the maxillary specimen known as _Meganthropus africanus_ because the occlusional surfaces showed completely different attrition from the P^3 and P^4 of the maxillary fragment. This separate M^3 has three roots and in its overall size it could have been part of the same maxilla the P^3 and P^4 belong to. Its length is 10.0 mm, and breadth 13.8 mm. These measurements are slightly higher than those of modern Australians and Bantus but below those of the Australopithecines of Swartkrans and Sterkfontein. They would be similar to Pithecanthropus IV.

Microanalysis of bone adhering to the roots of the molar, as well as its state of fossilization and mineralogical make-up of the tuff-deposit within the two fragments speak at least for a contemporaneity of the separately found molar and the maxillary fragments. Even though both hominid finds date chronologically to the same horizon and time period they can be assigned to two different hominids. Garusi I, the maxillary fragment with P^3 and P^4 belongs to an individual of fairly young age, possibly between 20 to 25 years old. Garusi II, the separate M^3, belongs undoubtedly to a different and much older individual since attrition is quite advanced. There is no possibility that such differentially worn teeth could belong to the same individual.

Weinert (1950) named the hominid _Meganthropus africanus_ since he thought it to be close morphologically to the Meganthropus of Java. This is a view not shared by most other specialists. Opinions range from assignation to Sinanthropus (Remane, 1951; 1954), a Preanthropine stage (Abel, 1940; Henning, 1948; Senyurek, 1955), _Plesianthropus transvaalensis_, and finally to the Australopithecines (Robinson, 1953). Most specialists agree that the fragment is hominid, even though most had the bad fortune of working with bad casts. To add to the problem, the alveoli of the canine and lateral and central incisors were covered by a tuffaceous matrix. Microanalysis on F, U, and N of the molar and the maxillary fragment show the same results. An amino-acid date was attempted but the fossils seem to be in age out of the range of this dating technique. Results of a date previously run of 800,000 years B.P. are misleading because of contamination of recent amino-acids due to preservatives.

Cleaning of the alveoli of the canine and the incisors provided a better knowledge of the size and morphology of these teeth. When reconstructed it appears that the canine projected above the occlusional level of the other teeth. It was quite large, substantially larger than canines of modern man. Apart from that, the matrix adhering to the maxillary fragment was compared to the mineralogical make-up of the Laetoli Beds and will hopefully be also compared to the matrix adhering to the new Laetolil hominids found by M. Leakey during the 1973-1976 excavations in the Garusi Area (Leakey et al. 1976; White, 1976).

The number of roots of P^3 and P^4 could finally be accurately determined by the use of x-ray analysis. It was found that P^3 had three roots, two buccal and one lingual, and P^4 two roots, one doubly fused buccal and one lingual root.

The morphology of P^3 and P^4 is not unlike that of modern Homo sapiens, only larger. P^4 is, contrary to earlier reports somewhat smaller than P^3, an essentially modern trait. Since P^3 has three roots it appears in this trait to be similar to some specimens from Swartkrans and also to Pongids. Mesial contact facets can be clearly recognized on P^3, P^4, as well as on M^3. They are all noticeable as concave areas on the upper part of the crown which is an indication of some crowding of the postcanine teeth. The contact facet on the mesial side of P^3 is located more buccally as is the case in most modern hominids. This again suggests that the contact zone of the canine is also more buccally located. It is, however, the small contact facet on P^3 which seems to indicate the presence of a fairly large canine. As mentioned above, a fairly large canine seems to have been present judging from the length of the alveolus as well as breadth and length of the alveolar margin. The root itself is much larger than in A. robustus or A. africanus. Finally, strictly judging from the minimum thickness of the interalveolar septi no diastema, neither precanine nor post-canine, seem to have been present. Even though only a small maxillary fragment is available a reconstruction of the dental arcade is possible. It is in every respect not unlike that of anatomically modern man, that is, the alveolar margin is curved convexly. The overall morphology and state of preservation of the maxillary fragment seems to be similar to the new Laetolil hominids (pers. comm. T. D. White, 1976). It is interesting to note that Senyurek as well as Oakley place the hominid into the Laetolil Beds (Oakley, 1968).

It seems beyond doubt that the individual belongs to some member of the Genus Homo. That it is a member of Homo erectus per seems to be somewhat doubtful. It seems instead plausible to include the individual in a group leading to Homo erectus. The inability of previous specialists to agree on one positive assignation seems to strengthen this argument.

According to the above the Garusi Hominids can be assigned to the Laetolil Beds which were recently re-excavated by M. Leakey (Leakey et al. 1976). Their similarity in morphology to these new hominids as well as absolute dating of them by K/Ar to somewhere between 3.5 and 3.77 million years B.P. makes it feasible to include them within that group.

It seems that neither Eyasi nor the Garusi hominids are positively members of "typical" Homo erectus in East Africa. If one searches for the closest similarity on an evolutionary scale the Garusi hominids could possibly be predecessors to Homo erectus. Until, however, a more detailed examination of new finds and further finds are available, only two hominid fossil specimens seem to definitely qualify for their inclusion in a Homo erectus group--that is in East Africa--namely Olduvai Hominid (OH 9) and most likely the Ndutu find.

Acknowledgments

I would like to thank H. Muller-Beck for the hominid material and Ms. Richter for the preparation of the manuscript as well as the Deutsche Forschungsgemeinschaft for the grants DFG Pr 143/1 and DFG Pr 143/3 which made a large amount of the research possible.

References

Bada, J.L. and Protsch, R. (1974). Concordance of Collagen-Based Radiocarbon and Aspartic-Acid Racemizations Ages. Proc. Nat. Acad. Sci. USA, Vol. 71, No. 3, p. 916.

Birkner, F. (1937). Reste des Urmenschen in Afrika? Germania.

Clark, R.J. (1976). New Cranium of Homo erectus from Lake Ndutu, Tanzania. Nature Vol. 262, p. 485-487.

Hennig, E. (1948). Quartärfaunen und Urgeschichte Ostafrikas. Naturwiss. Rdsch., I, pp. 212-217.

Kohl-Larsen, L. (1943). Auf den Spuren des Vormenschen. Forschungen, Fahrten und Erlebnisse in Deutsch-Ostafrika. Band I und II, Strecker und Schroder Verlag, Stuttgart.

Leakey, L.S.B. (1936). A New Fossil Skull from Eyasi, East AFrica. Nature, Vol. 128, pp. 1082-1084.

Leakey, L.S.B. (1946). Report on a Visit to the Site of the Eyasi Skull, Found by Dr. Kohl-Larsen, Journal of East Africa nat. Hist., Soc., 19, pp. 40-43.

Leakey, L.S.B. (1948). Fossil and Sub-Fossil Hominidea in East Africa. Robert Broom Commemorative Volume 1948. Spec. Publ. Royal Soc. South Africa, pp. 165-170.

Leakey, L.S.B. (1952). The Age of the Eyasi Skull. Proc. Pan-Afr. Congress on Prehist. 1947, Nairobi. Oxford U. Press.

Mturi, A.A. (1976). New Hominid from Lake Ndutu, Tanzania. Nature Vol. 262, p. 484-485.

Oakley, K.P. (1968). Frameworks for Dating Fossil Man. Chicago. Aldine Publ. Co., pp. 1-8.

Protsch, R. (1975). The Absolute Dating of Upper Pleistocene SubSaharan Fossil Hominids and Their Place in Human Evolution. Journal of Human Evolution 4, pp. 297-322.

Protsch, R. (1976). The position of the Eyasi and Garusi Hominids in East Africa. Colloque VI,

Les Plus Anciens Hominide's, Centre National de la Recherche Scientifique, IXe Congrès Union Internationale des Sciences Prehistoriques, Nice 1976.

Protsch, R. (1977). A New Morphological analysis based on a reconstruction and dating of the Garusi and Eyasi hominids. Kohl-Larsen Festschrift, Band II, ed. H. Muller-Beck.

Reck, H. and Kohl-Larsen, L. (1936). Erster Überblick über die jungdiluvialen Tier- und Menschenfunde Dr. Kohl-Larsen's im nordostlichen Teil des Njarasa-Grabens (Ostafrika). Geologische Rundschau, Sonderdruck, Vol. 27, 5, pp. 401-441

Reeve, W.H. (1946). Geological Report on the Site of Dr. Kohl-Larsen's Discovery of a Fossil Human Skull, Lake Eyasi, Tanganyika Territory. Journal of East Africa nat. Hist. Soc., 19, pp. 44-50.

Remane, A. (1951). Die Zähne des Meganthropus africanus. Zeitschrift für Morphologie und Anthropologie, Bd. XLII, Heft 3, pp. 311-329.

Remane, A. (1954). Structure and relationships of Meganthropus africanus. American Journal of Physical Anthropology, Vol. 12, New Series, No. 1, pp. 123-126.

Robinson, J.T. (1953). Meganthropus, Australopithecines and Hominids. American Journal of Physical Anthropology, Vol. 11, New Series, No. 1, pp. 1-38.

Senyürek, M. (1955). A note on the Teeth of Meganthropus africanus Weinert from Tanganyika Territory. Belleten, XIX, 73, pp. 1-57.

Weinert, H. (1937). Hominidae (Paläozoologie). Fortschritte der Palaontologie, I, pp. 337-344.

Weinert, H. (1938). Der Neue Affenmensch "Africanthropus". Germania, I. pp. 21-24.

Weinert, H. (1938a). Africanthropus, Der erste Affenmenschen-Fund aus dem Quartar Deutsch-Ostafrikas. Quartar, 1, pp. 177-179.

Weinert, H. (1938b). Africanthropus (Pithecanthropus Stufe), in: Entstehung der Menschenrassen, Verlag F. Enke, Stuttgart, pp. 33-40.

Weinert, H. (1938c). Der erste afrikanische Affenmensch "Africanthropus njarasensis". Der Biologe, Vol. 4, pp. 125-129.

Weinert, H. (1940). Africanthropus, der neue Affenmenschenfund vom Njarasa-See in Ostafrika. Zeit. Morph. u. Anthrop. 38, Vol. 1, pp. 18-24.

Weinert, H. (1950). Über die neuen Vor-und Fruhmenschenfunde aus Afrika, Java, China und Frankreich. Zeit. Morph. u. Anthrop. XLII, 1, pp. 113-148.

Weinert, H. (1952). Über die Vielgestaltigkeit der Summoprimaten vor der Menschwerdung. Zeit. Morph. Anthrop., 43, pp. 73-103.

Weinert, H., Bauermeister, W. & Remane, A. (1940). Africanthropus njarasensis. Beschreibung und phylethische Einordnung des ersten Affenmenschen aus Ostafrika. Zeit. Morph. u. Anthrop. 38, Vol. 1, pp. 252-308.

White, T.D. in Leakey, M.D. et al (1976). Fossil hominids from the Laetolil Beds. Nature Vol. 262, p. 460-466.

THE STRUCTURE OF ANATOMICAL FRAGMENTS, AND THEIR COMBINATION: A PROBLEM UNDERLYING THE ASSESSMENT OF THE AUSTRALOPITHECINES

Charles E. Oxnard
University of Southern California

The Evolution of the Primates

Many chapters in this book deal directly with the australopithecine fossils and associated data relating to their geology, environment and artefacts, and the history of our study of them. But a few chapters apply themselves to the australopithecine story indirectly through background investigations of living primates. Such a background is as indispensible in understanding these species as are the primary materials themselves. It is this latter approach that I would like to continue in this chapter.

The general plan of the relationships of the living primates has been settled for a long time now, although it is true that individual points are challenged and changed from time to time. That plan stems in part from a pattern of similarities that sees the living primates as a relatively linear sequence with the prosimians lying at one extreme, the New World monkeys next, then the monkeys of the Old World, and the apes and humans at the other extreme. It is to W.E. Le Gros Clark (1959) that we can turn for a diagram (figure 1) that summarizes these arrangements. It supplies an assessment about systematic and evolutionary relationships wherein the great bulk of the information is derived from morphology. This includes information, first, from bones especially teeth, jaws and various anatomical regions of the cranium, and, to much lesser degrees, the post-cranium and, second, from external features such as the color and pattern of the pelage, the form of the face and genitalia, the characteristics of the hands and feet. That these should be the primary data resulting in our view of the primates is not surprising - such data have been known for decades, indeed centuries.

Other types of information are also available to help provide assessments about relationships, about evolution. Thus physiological parameters, especially those relating to reproduction, development and growth,

FIGURE 1: Arrangement of the primates as derived from classical morphological data.

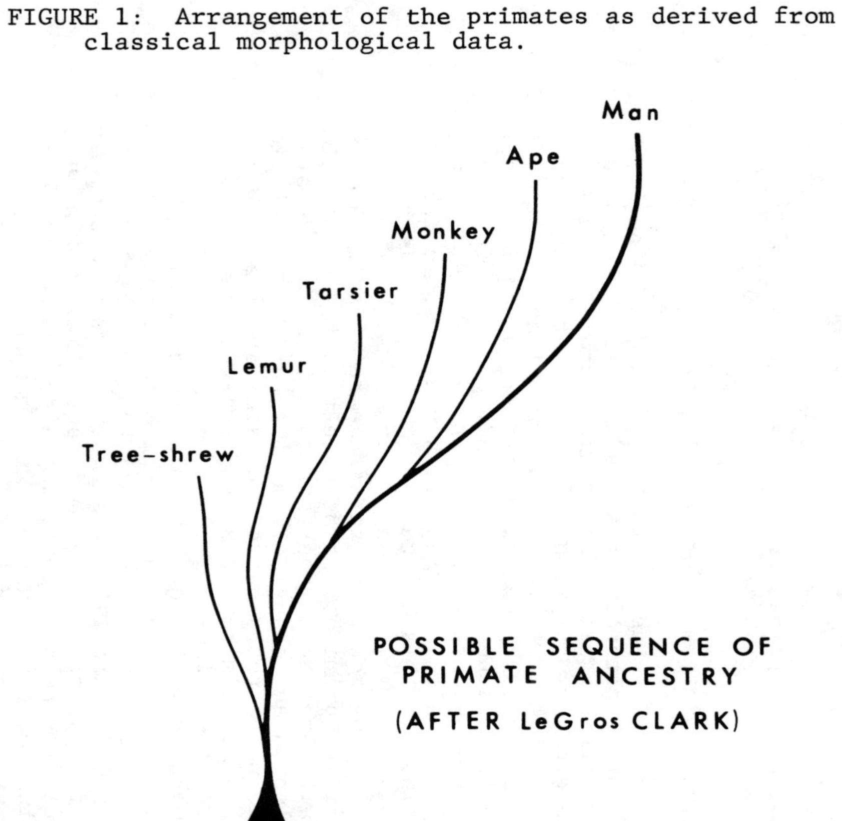

bring important facts to bear upon these problems. It is to A.H. Schultz (1969) that we may turn for a picture that summarizes these findings. Figure 2 shows how the primates are grouped when we use, as the defining characteristic, a physiologically based datum such as growth after birth. Again, a linear spectrum is obvious with prosimians at one end, New World monkeys, Old World monkeys and apes following them in sequence, and with humans at the other, opposite, extreme. Data of this sort have also been available for many years.

Finally, the relationships of animals can be defined through examination of cellular, biochemical and molecular parameters. This notion is far older than often surmized; the oldest studies date from the turn of our century (e.g. Friedenthal, 1900; Nuttall, 1904) and a review written as early as the third decade spelt out the possibilities in remarkable detail (Zuckerman, 1933). Even much earlier, however, Darwin himself had a clear understanding of the general principle. However, it has only been in the last twenty years that a sufficient bulk of such data has become available to make assessments for the entire Order. And although there are many different investigators and methods, and just as much controversy and disagreement here as for other kinds of data, nevertheless a broad consensus can be demonstrated, as an example, the work of M. Goodman (1976) summarized in figure 3. This shows some relationships among living primates based upon examination of antigenetic distances. It resembles information stemming from other types of biochemical and related studies such as blood group distributions, chromosome arrangements, protein sequences, immunological tolerances, DNA hybridisations, and so on. And the picture that is portrayed is essentially the same as for the morphological and physiological approaches: the species are arranged in linear sequence - prosimians, New World monkeys, Old World monkeys, hominoids; in particular humans lie again at the opposite end to prosimians.

Whatever, then, may be the points of controversy in primate systematics and evolution, the basic arrangement has long been known and is now established by a series of different approaches (see summary in Oxnard, 1982). When, however, it comes to judging fossil data the above methods are considerably less viable. Of the classical morphological studies really only a few, comparisons of teeth, jaws and fragments

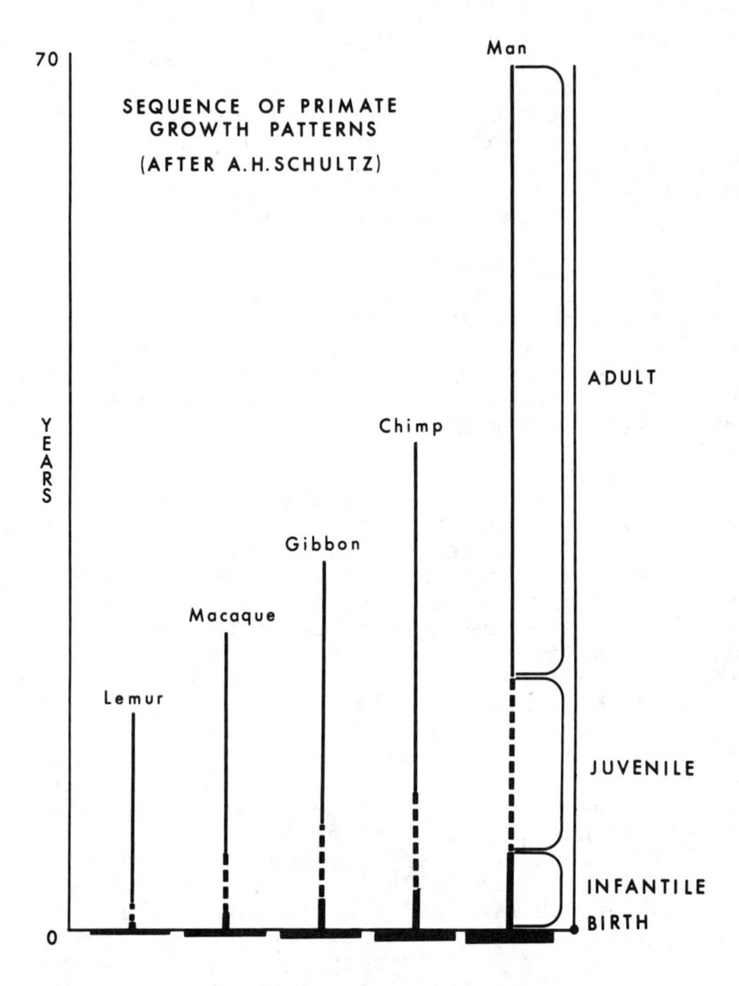

FIGURE 2: Arrangement of the primates as derived from study of postnatal growth patterns.

FIGURE 3: Arrangement of the primates as derived from antigenic sequences. The plot has been rotated for comparison with figures 1 and 2.

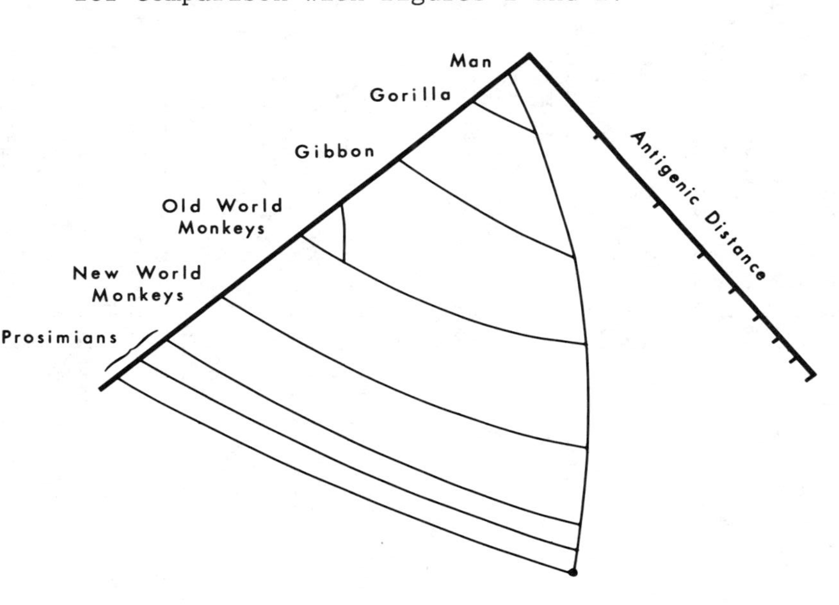

of crania, are available for the fossils; post-cranial materials have been, in the past, most uncommon although this is now slowly changing with the new fossil finds. Little at all can come from external features save what can be elaborated from study of the facial skeleton. Of the physiological investigations only the barest inferences can be made from secondary implications about bone fragments. And, of course, it is in only very rare instances that it is possible to obtain direct biomolecular or biochemical information.

Furthermore, for fossils, data must usually, in the first instance, be viewed piecemeal. Only after study of the initial fragments may we attempt to put fossil pieces together because of our uncertainties about whether or not they really do go together. Individual variation, differences in growth patterns, the complexities of sexual dimorphism, the possibilities of geographic and other varieties, may all confuse the issue enormously.

And, finally, for fossils, there are different ways of evaluating the data. Whereas for the teeth, jaws, and cranium, most anatomical information has been used to arrive directly at phylogenies, in the post-cranium information rarely points immediately to phylogenetic arrangements. In the post-cranial skeleton, anatomy usually bespeaks function; and it is only some notion about the evolution of function that can secondarily help us with the evolution and systematics of the animals. It is thus necessary to try to understand the functional relationships of living creatures as revealed by post-cranial anatomy and osteology before we can define what the post-cranial anatomy of the fossils may be telling us. _Sotto voce_, I am sure that this should also be the case for the cranium, but I suppose that structural functional relationships in the many regions of the skull are so little understood that most earlier investigators have made the direct leap to phylogeny.

The Structure of Upper Limbs

Our earliest essay into this particular realm of primate functional anatomy (summarized in Oxnard, 1973a) was an attempt to understand something of the structure of the shoulder. It was done not only through associating elements of behavior (such as locomotion) with soft tissue anatomy (muscles and joints) and hard tissue conformations (bone and cartilage), but also by defining the form of bones through the methods of biometrics and multivariate statistical analysis. A summary of that work is provided in figure 4 and at first sight it might be thought that it agrees with the information in figures 1 through 3; certainly the primates are arranged in a linear sequence.

But as we study the precise locations of individual genera within that sequence it becomes apparent that it differs totally from that presented above. Prosimians are located at many places throughout the sequence, as are both New World and Old World monkeys. Only apes are confined to one end of the spectrum; and the position of those other hominoids, humans, is not near apes, but central in this particular view of the results. It is apparent, then, that study of the shoulder alone does not give information of immediate relevance to the evolution of the primates. What then does the shoulder arrangement imply?

If we restudy the placement of the various genera with some notions about the function of the shoulder in mind, an answer begins to suggest itself.

Thus, those creatures (of whatever taxonomic group) that are placed at the lower end of the spectrum resemble one another in the sense that, during the various behaviors, especially locomotor behaviors, of which they are capable, their shoulders are used most frequently in dependent positions and bearing compressive forces. They include forms such as baboons and patas monkeys, squirrel and owl monkeys, lemurs and bush-babies.

The creatures towards the other end of the spectrum, irrespective of their taxonomic placement, resemble one another in the sense that, during the various behaviors, especially locomotion, of which they are capable, their shoulders are used more often in raised positions and bearing tensile forces. (But of course, the actual bones of the shoulder do not habitually bear tension - they presumably are under compression as a result of tension in the soft tissue ties: muscles and ligaments). This is so for species such as gibbons and siamangs, orangutans and chimpanzees, spider monkeys and woolly spider monkeys.

Centrally located within the sequence are a variety of forms which, whatever may be their overall behaviors, function in such a manner in climbing and foraging, that their shoulders execute movements and bear forces that are intermediate between those in the shoulders of the two sets of species just described. This is so whether

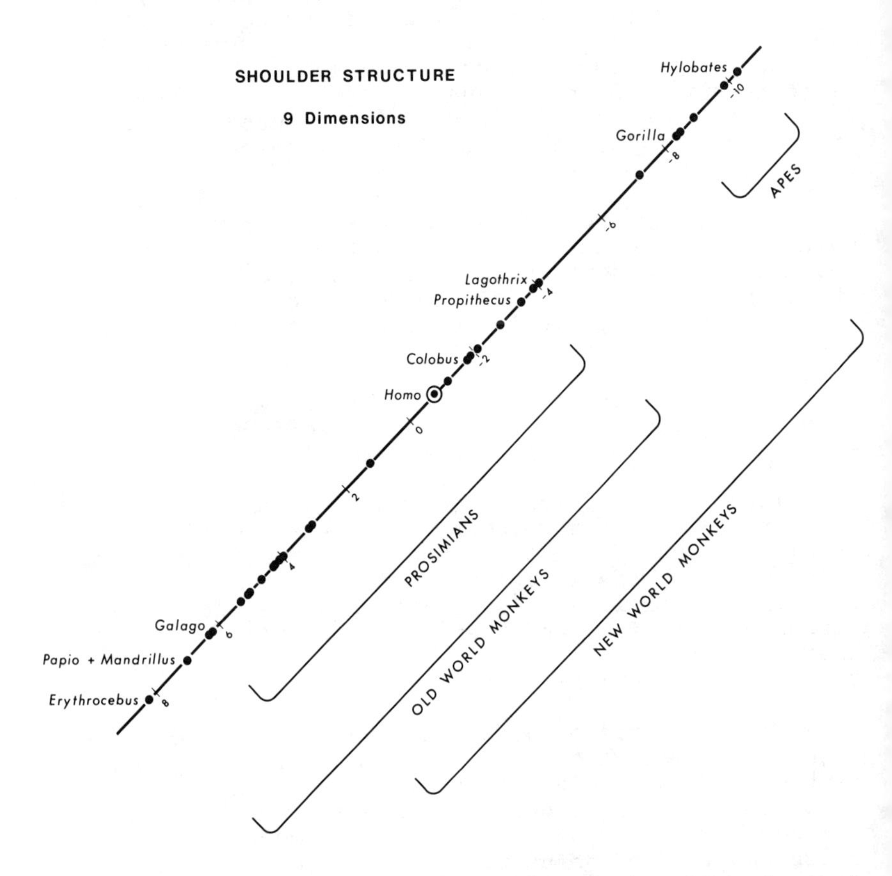

FIGURE 4: Arrangement of the primates by the principal canonical axis derived from multivariate statistical studies of the primate shoulder. The brackets indicate the overlap between major taxonomic groups. Units are in standard deviations.

the animals display the kinds of activities found in colobus monkeys and many langurs, or those found in woolly monkeys and to lesser degrees in howler monkeys, or even those displayed in the vastly different locomotion of prosimian species such as pottos and angwantibos, indris and sifakas.

The totality of this picture is, then, that study of the structure of the shoulder by such methods is far more able to predict the function of the shoulder than to tell about the evolution of the animals. Such a conclusion can only be certain however, if it can be confirmed by study of other anatomical regions. And because the shoulder is a bone so completely suspended by muscles, its shape therefore so completely determined by muscular forces, we would not be overly surprised if other anatomical regions did not present such a clear functional picture.

Equivalent studies have therefore been carried out upon the arm, forearm and forelimb as a whole (Ashton, Flinn, and Oxnard, 1975; Ashton, Flinn, Oxnard and Spence, 1976; Oxnard, 1975a and b). It turns out that although it seems that the shoulder is indeed most sensitive to the impress of function, so also are other anatomical regions. The arrangement of the animals found in the shoulder studies is paralleled in studies of these other regions of the upper limb. Thus, figure 5 demonstrates, unequivocally, that different investigations of the shoulder, arm and forearm all provide approximately similar linear arrangements of the primates. And figure 6 demonstrates, equally unequivocally, that study of the overall proportions of the entire forelimb (taken by the late Professor A.H. Schultz who kindly made them available to me in his lifetime) produces the same picture. In this case the data are best seen through a particular two-dimensional view (of a three-dimensional model) rather than a one-dimensional axis. The morphological arrangement parallels function with a mixture of taxonomic groups, rather than evolution with a linear sequence of prosimians, monkeys, apes and humans.

A final method of investigating whether or not the information contained within these morphological arrangements is truly associated with behavioral realities is to investigate those species contained within the dotted circle of figure 6. These species all seem to fall close together morphologically, and they all display locomotor modes that have been grouped

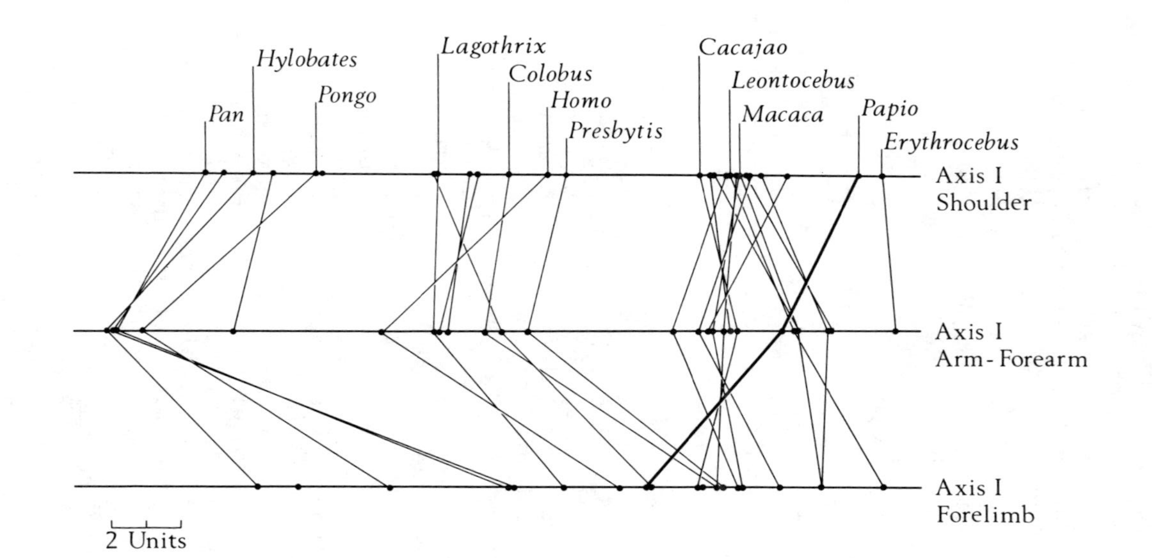

FIGURE 5: Comparison of rank orders of genera in the principal canonical axes in each of three studies--shoulder, arm and forearm, and upper limb as a whole. Units in standard deviations.

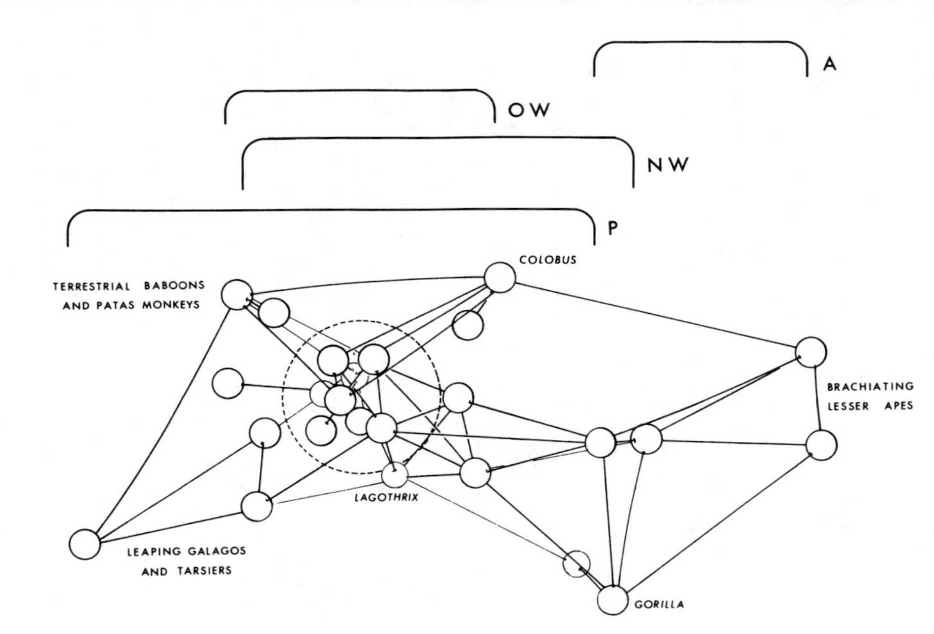

FIGURE 6: The three-dimensional model that can be built from the generalized distances separating primate genera in the analysis of overall proportions of the upper limb. Brackets indicate overlap in taxonomic groups. The regular arboreal quadrupeds lie close to or within the dotted circle. General scale of the model = 30 distance units in length.

as generalized arboreal quadrupedalism. In fact, however, there are definable differences among them.

Some are less arboreal (e.g. *Cercopithecus aethiops*, *Macaca irus*, *Presbytis entellus*); others are by comparison, somewhat more arboreal (e.g. *C. diana*, *M. nemestrina*, *P. obscurus*). Some are considerably less acrobatic in their use of their forelimbs in the trees (e.g. marmosets and titi monkeys); others are considerably more acrobatic in upper limb usage within the trees (e.g. uakaris and woolly monkeys). An attempt (Oxnard, 1978a) has been made to investigate the minor differences in structure between these various quadrupedal New and Old World monkeys. Figure 7 demonstrates that the species fall into mini-spectra as indicated by the sequence of envelopes in the diagrams. And these mini-spectra relate not to similarities in taxonomic group (i.e. not to all *Cercopithecus* species together, or all cebid genera together), but rather to similarities in function (i.e. all lesser arboreal species together as compared with all more arboreal ones, and all lesser acrobatic genera together as compared with all more acrobatic ones). In each case, in both Old World monkeys and New World monkeys but separately within each, the envelopes of quadrupedal forms are arranged in such a way as to reflect functional differences.

Thus the envelopes labelled A contain those Old World species that are least highly arboreal and those New World genera that, while being totally arboreal, are least highly acrobatic in the trees. The envelopes labelled C contain those Old World monkeys that are most highly arboreal, and those New World forms that are most acrobatic in the trees (save that the woolly-spider and spider monkeys are excluded from the analysis because the degree of their acrobatic activity is such that they can scarcely be thought of as regular quadrupeds-the initial defining criterion). The envelopes labelled B contain, in each case, those species and genera that are intermediate in arboreality and acrobatic activity, respectively. Finally, the relationships between the sets of envelopes, irrespective of the individual tortuosities of the different envelopes, is such that envelope B lies always between envelopes A and C. Even at this very detailed level of morphological assessment, the behavioral associations hold up remarkably well.

Early pilot studies at the species and subspecies levels (Oxnard, 1967) suggested that the subtleties of locomotor adaptations of structure are very fine indeed.

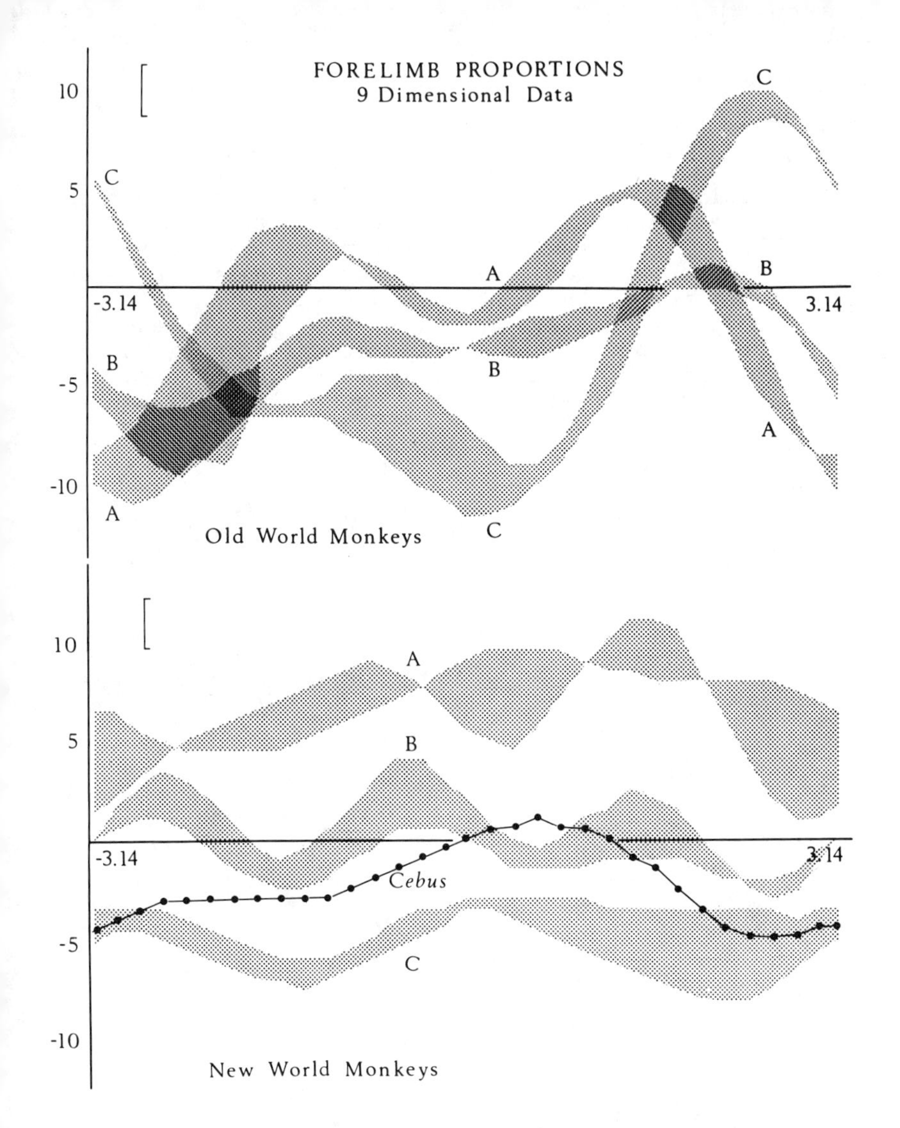

FIGURE 7: Analysis using Andrews' high dimensional display (Andrews, 1972, 1973; Oxnard, 1973, 1975a) to reveal relationships among generally quadrupedal Old World and New World monkeys. Each individual genus is represented by a curve plotted from -pi to +pi. Similarity of curves means morphological similarity; differences of curves means morphological differences. In this figure, those genera that are closely similar have been

Legend continued:

coalesced into shaded envelopes. The method of display retains the concept of statistical variance and a single standard deviation unit marker is shown. The area of an envelope is the generalized distance between similar genera; the area between envelopes is the generalized distance between groups of genera. Groups A, B, and C for each of the New World and Old World groups of monkeys are, as explained in more detail in the text, defined from the activities of the upper limbs in locomotion.

Manaster (1975, 1979) has taken those ideas much further and has shown precise morphometric differences among several parallel sets of species of cercopitheques, mangabeys and langurs that have parallel behavioral (locomotor) variations. Fleagle (1976) has confirmed this in two of the species of the langurs, Mittermeier (1978) has extended the idea to two species of Ateles, and Rodman, (1979) to two species of macaque. McArdle (1978, 1981), Oxnard, German and McArdle (1981) and Oxnard, German, Jouffroy and Lessertisseur (1981) have demonstrated similar parallels in considerable detail for several prosimians (e.g. lorises, lemurs, mouse lemurs, bush-babies).

An initial reaction to these results might be to question them because they are so different from the overall picture presented by the studies summarized in figures 1 through 3. However, a number of other investigators have either replicated these studies or carried out investigations that are basically the same. Thus, there is a marked similarity between our studies of the arm and forearm (Ashton, Flinn, Oxnard and Spence, 1976) and those of Feldesman (1976) who was doing a parallel study at the same time (figure 8); and there is a marked similarity (Oxnard, 1977a) between the appropriate part of our original studies of the shoulder girdle (Ashton, Flinn, Oxnard and Spence, 1971) as compared with a more restricted investigation carried out by Corruccini and Ciochon (1976; see figure 9).

Another aspect of the results, the position of humans, may, initially, also appear to deny the functional interpretations that I have offered. Thus, in each case, in the studies of the shoulder, arm, forearm and overall upper limb proportions, the genus Homo appears to be centrally placed within the matrix of non-human species. Such a placement would predict a functional role for the human upper limb very similar to that of any number of non-human primates. And this is not, of course, the case.

In fact, this apparent result is an artefact of using overly simple one-dimensional (figures 4 and 5) and two-dimensional (figure 6) displays of more complex results. In each study it is easily shown that humans actually lie in positions markedly off-set from those of non-human primates, usually in one or more dimensions higher than those represented here. This is illustrated for the study of upper limb proportions by

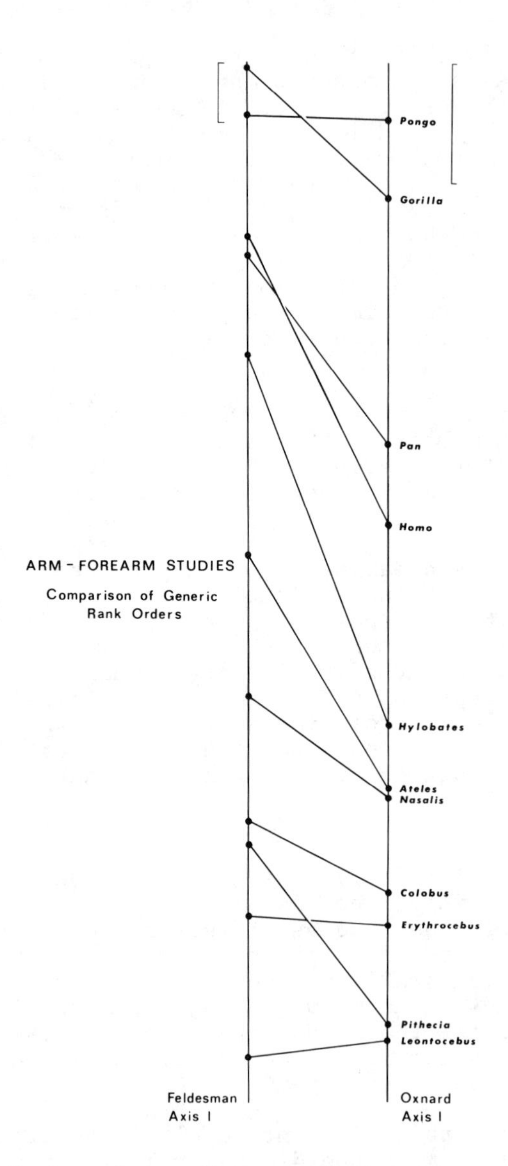

FIGURE 8: Rank orders of the primate genera in the first canonical axis of two studies of the arm and forearm by two independent investigators as indicated. The two studies are almost identical. The single standard deviation marker pertains to both studies.

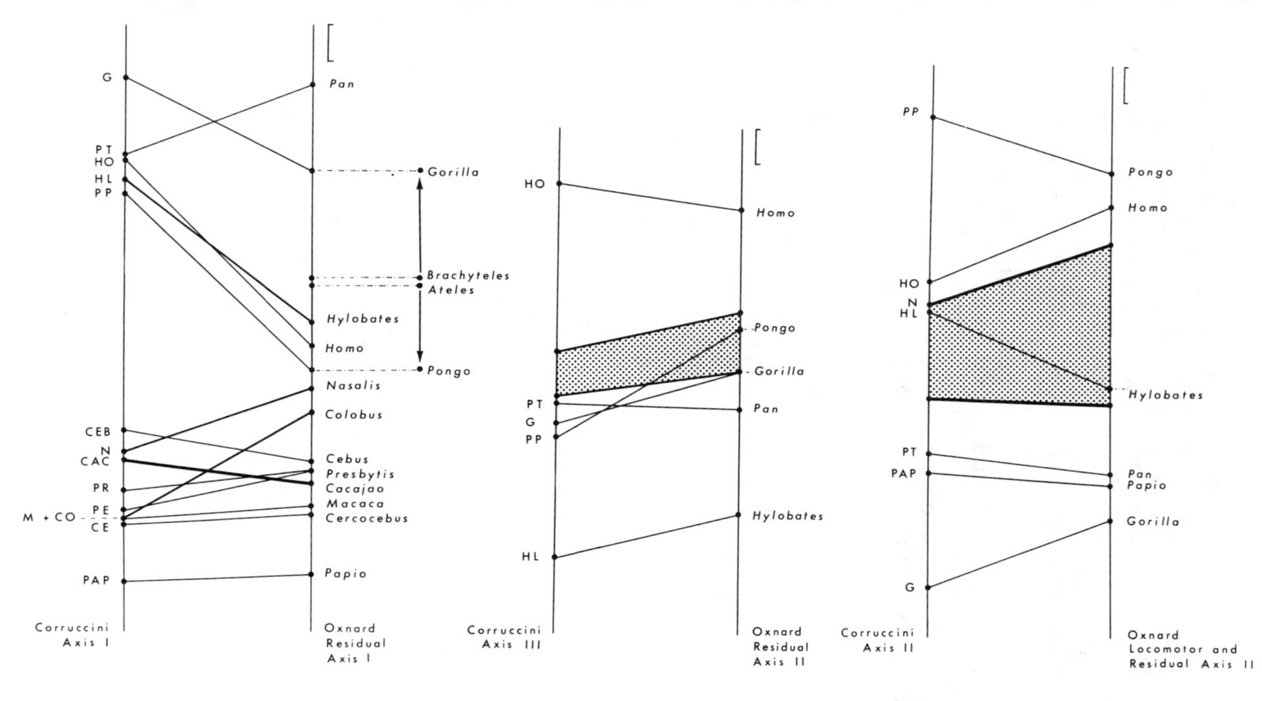

FIGURE 9: Rank orders of the primate genera in each of the first two axes of two separate studies of the primate shoulder girdle by different investigators as indicated. There are slight differences

Legend continued:

between the first pairs of axes that relate to the existence of two extra genera (_Ateles_ and _Brachyteles_) in the analysis by Oxnard. There is also a biologically unimportant difference in that the information in axes two and three is reversed in the two studies. The shaded areas contain many genera of monkeys not appreciably separated by these axes yet in approximately the same positions in each. The scale marker is one standard deviation for each axis of the canonical variates analyses of Oxnard and colleagues; no markers are available for the principal coordinate axes of Corruccini and Ciochon. The results of these two studies are remarkably similar.

rotating the model shown in figure 6 to allow the third dimension to be seen (figure 10). Humans are obviously not centrally located with reference to the structure of the upper limb in non-human primates; they are uniquely different. This presumably confirms the functional explanation of the results because humans have upper limb functions unique among the primates.

This unique separation of humans could be of great import in examining fossils presumed to be on the pathway of human evolution. For there must have been some time in the past when an ancestor of the human species was in the process of changing, functionally, from some non-human usage of its upper limbs towards a human-like usage. The position of that ancestor in relation to both humans and the entire spectrum of living non-human primates might provide powerful clues as to how, functionally, this upper limb change had occurred.

The Structure of Lower Limbs

The arguments posited above for upper limbs are sound only if they can be confirmed in other anatomical areas and especially if they can be confirmed in lower limbs. Here a lesser range of investigations are available: studies of the pelvis (Zuckerman, Ashton, Flinn, Oxnard and Spence, 1973; Oxnard, 1973a,b; Ashton, Flinn, Moore, Oxnard and Spence, 1981), the talus (Lisowski, Albrecht and Oxnard, 1974, 1976), the entire foot (Oxnard, 1980, Lisowski and Oxnard, 1980) and the overall proportions of the lower limbs (Ashton, Flinn, and Oxnard, 1975; Oxnard, 1973b, 1979a,b).

Again, the first essay was into functional anatomy; an attempt was made to understand pelvic morphology through correlation of soft tissue anatomy (muscles, ligaments and joints) with pelvic function during behavior, mainly locomotion. And, again, the osteological element of the investigation was carried out not only through observing the conformation of the bones to the positions and orientations of the various soft tissue structures, but also through the biometric definition of bone form obtained using measurement and multivariate statistical analysis. Studies of the talus were less based upon prior dissection and more upon what was already known about the soft tissues relating to the talus and its function as evidenced in the literature. But studies of the talus conformed to the above pattern in that its form, too, was defined not only visually, but also biometrically and using multivariate statistical analysis. Studies of the

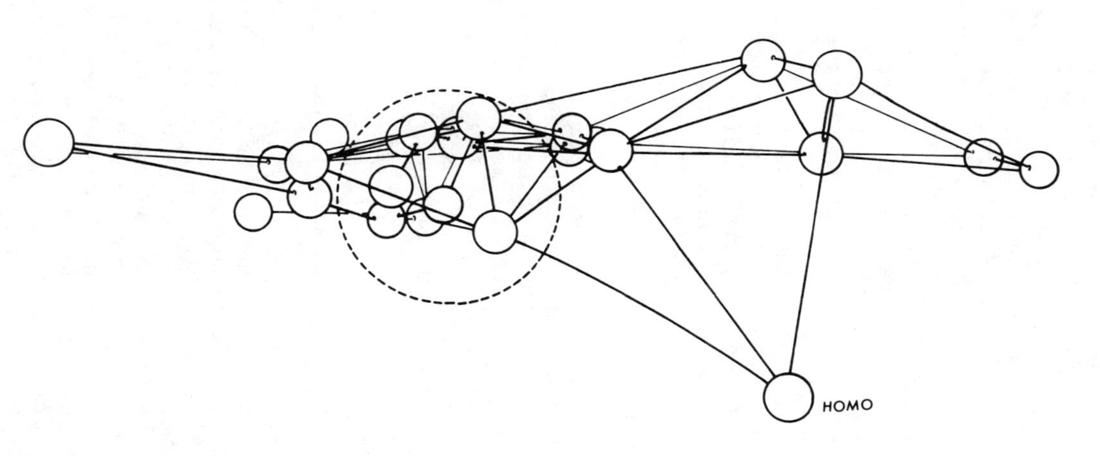

FIGURE 10: The three-dimensional model of upper limb proportions shown in figure 6 after rotation through 90 degrees around a horizontal axis demonstrates the uniqueness of humans. This is also the case for most other studies of upper limbs.

entire foot rested heavily upon prior dissection, but not, so far, upon biometry and analysis at all. But it is to the investigations of overall lower limb proportions (again the data of Professor Schultz studied by multivariate statistical analysis) that we may turn for a single picture that summarizes all these results.

Thus, figure 11 demonstrates the arrangement of the different primate genera in the study of lower limb proportions. What is first obvious is that this result in no way resembles that produced by the three approaches outlined in figures 1 through 3 in the introduction to this chapter. Rather than a linear spectrum, the arrangement of forms is star-shaped with many genera embedded in the center of the star (encircled in the diagram) and with many others lying in different rays of the star. Such an arrangement does not at all replicate evolutionary relationships. But when we come to note just where each primate genus is located in this structural result we can see that the associations are obviously functional.

The centrally located genera include all those that would normally be recognized as regular arboreal quadrupeds irrespective of taxonomic group. Thus lemurs, marmosets, owl monkeys, macaques, and many cercopitheques among others share this central locus.

The genera within peripheral rays of the star are similar only in that they have similar functions of the lower limb within locomotion. Thus the baboon and the patas monkey both share one ray, and both are terrestrial quadrupeds, even though each is evolutionarily closer to genera within the center of the star (patas monkeys to cercopitheques, baboons to macaques and mangabeys). Tarsiers and bush-babies share another ray of the star and both are extreme leaping animals; but bush-babies are more closely related to other prosimians within the center of the star, while the evolutionary relationships of tarsiers are more likely with the centrally located members of the anthropoidea. The spider and woolly spider monkeys of the New World fall in a ray of the star together with the gibbons and siamangs of the Old World. This similarity makes sense in terms of the functions of the lower limb within the acrobatic arboreal activities of which both are capable. Each is, however, related in an evolutionary sense to other species located more centrally in the nucleus of the star, than to each other.

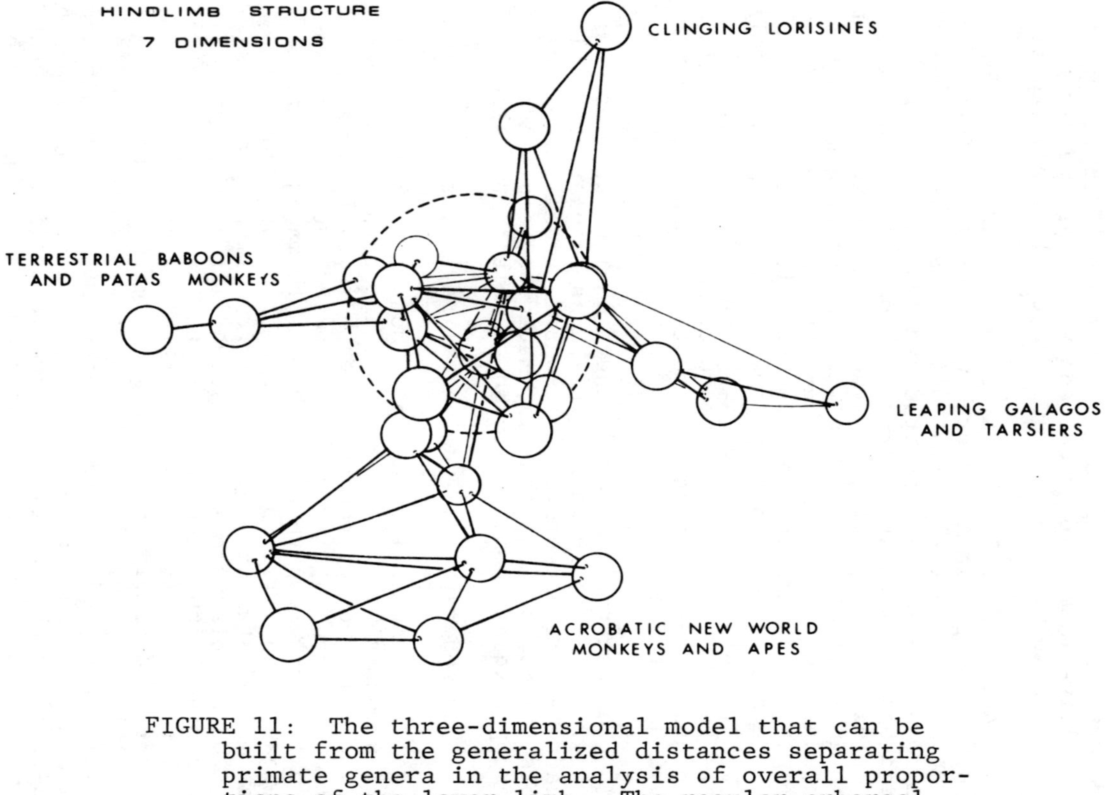

FIGURE 11: The three-dimensional model that can be built from the generalized distances separating primate genera in the analysis of overall proportions of the lower limb. The regular arboreal

Legend continued:

quadrupeds lie close to or within the dotted circle. General scale of the model = 35 distance units in length. In this case the result is more complex than for figure 6 so that the model is only locally three-dimensional. The result of this is that relationships between close genera in the model are correct; those among distant genera are longer than shown. Thus the model is the three-dimensional reduction of a higher-dimensional reality. Nevertheless, this model provides a picture of the relationships among the genera that is not excessively distorted and does not conceal major information. Thus the model is generally star-shaped with rays that are somewhat twisted. Neighboring genera in the nucleus and rays of the star are not related taxonomically, but possess parallels of locomotor function in their lower limbs.

This view of these functionally extreme parallels among the primates seems to confirm an association between lower limb function and lower limb structure. But, as with the upper limb studies, it is worth realizing just how sensitive this mode of analysis is to even subtle functional adaptations. If we study all those species located in the nucleus of the star (the regular arboreal quadrupeds contained within the dotted circle of figure 11) we find, as with the upper limb, that the various genera can be further separated on functional grounds. Thus figure 12 shows that, among quadrupedal Old World monkeys, species are separated according to greater or lesser degrees of arboreality. Figure 12 likewise demonstrates that among New World monkeys which are all fully arboreal, genera are grouped according to greater or lesser degrees of acrobatic activity within the trees. And figure 12 notes that, in each case, intermediate functional descriptions lie with intermediate morphological positions. This is, then, yet further evidence that what we are seeing here in these morphological arrangements are patterns that are associated with the function of the parts within locomotion.

Again, an initial reaction might be to question these results because they differ so greatly from the arrangements discussed at the beginning of this chapter. But again, some (though fewer) replicate studies are available for comparison (Day and Wood, 1968; McHenry and Corruccini, 1975). Each independent result parallels those described in our studies (Oxnard, 1977a; figures 13 and 14).

Again, too, the positions of humans might be thought to negate the functional ideas. However, as with upper limbs, study of all aspects of the results makes it clear that humans actually lie projected uniquely away from the locus of any non-human primate (for example, see figure 15). This presumably relates to the unique functional arrangement in humans; not only are they the only primates that are totally capable bipedally, they are also the only ones that are totally incapable quadrupedally.

And, again, the existence of this unique human separation could be of great value in attempts to determine the particular functional pathway taken during the evolution of lower limb elements in pre-human fossils.

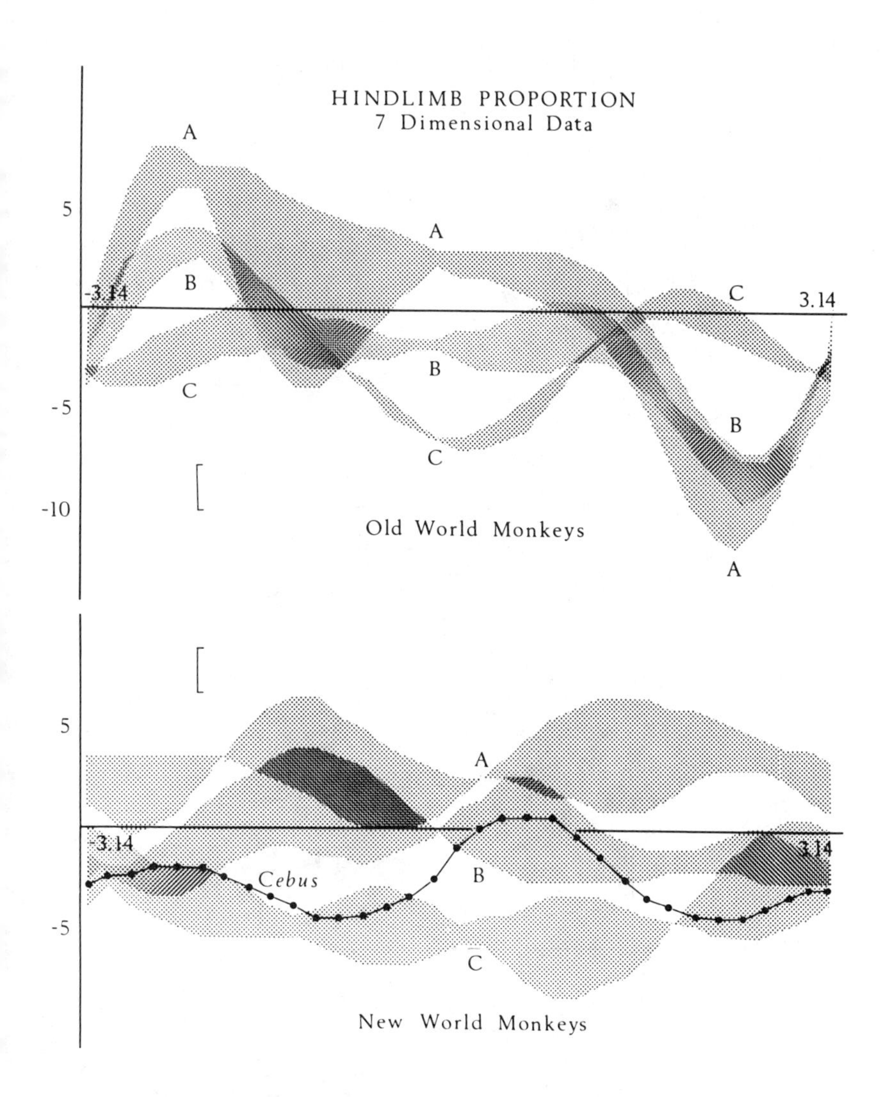

FIGURE 12: Analysis, in the same mode as figure 7, of the relationships of lower limbs among those New and Old World genera that are generally considered to be regular quadrupedal species and which were found to occupy the dotted circle in figure 11. Groups A, B, and C for each of the New World and Old World groups of monkeys are, as explained in more detail in the text, defined from the activities of the lower limbs in locomotion.

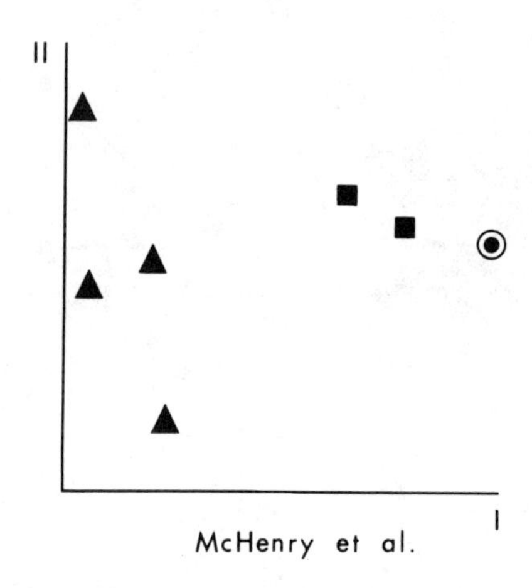

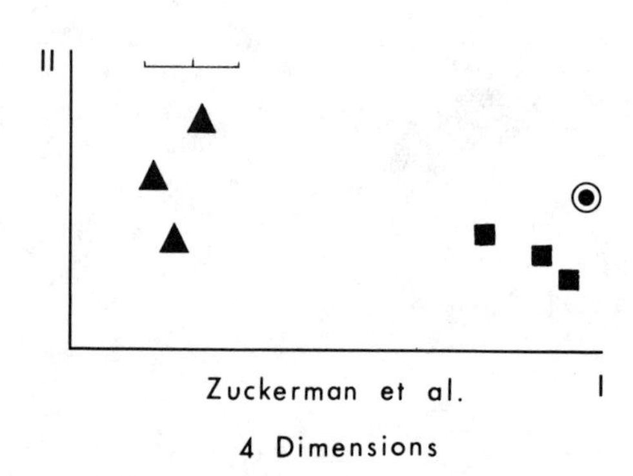

FIGURE 13: Bivariate plots of the first two axes in studies of the primate pelvis from the works of two separate groups of investigators as indicated. There is an extra genus (a lesser ape) in the one study not represented in the other. Triangles = apes, circle = humans, square = fossil. A single standard deviation unit is available for the study of Oxnard. The similarity between the two studies is most marked.

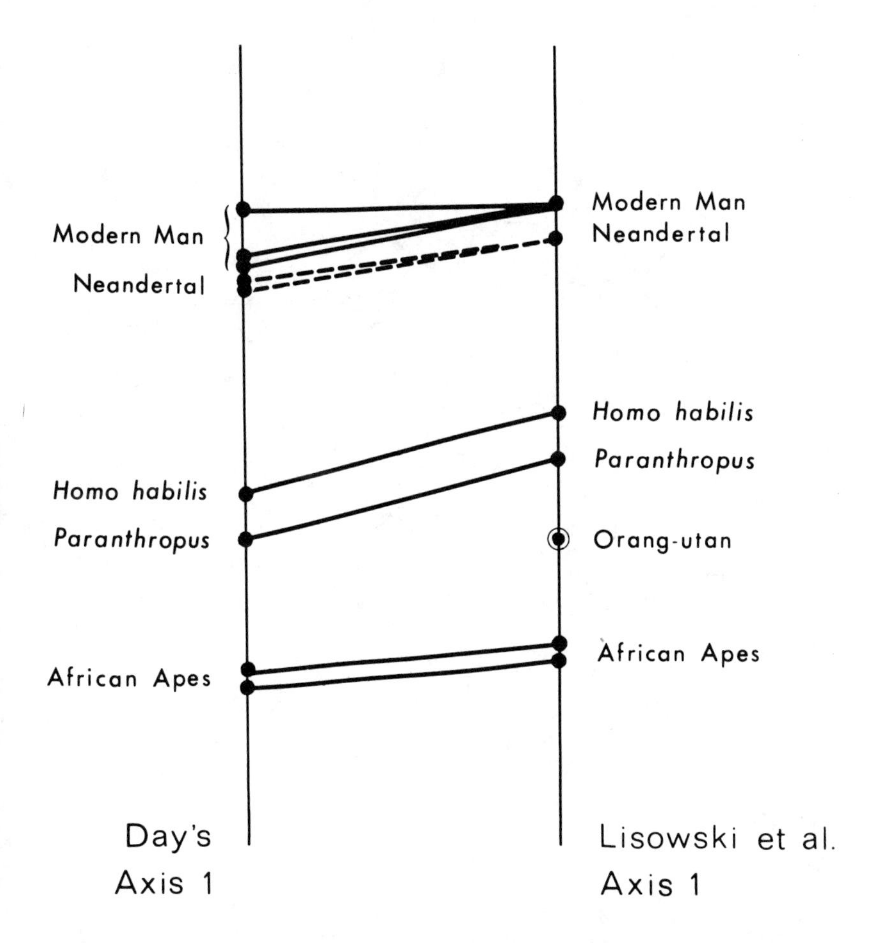

FIGURE 14: Rank order of primate genera in the first axis in studies of the primate talus from the works of two separate investigators as indicated. One extra genus, Pongo, is available for the studies of Oxnard and colleagues. The separation between humans and African apes is about eight standard deviation units in both studies. The similarity between the results is most marked although interpretations differ because of the absence of Pongo in the second study.

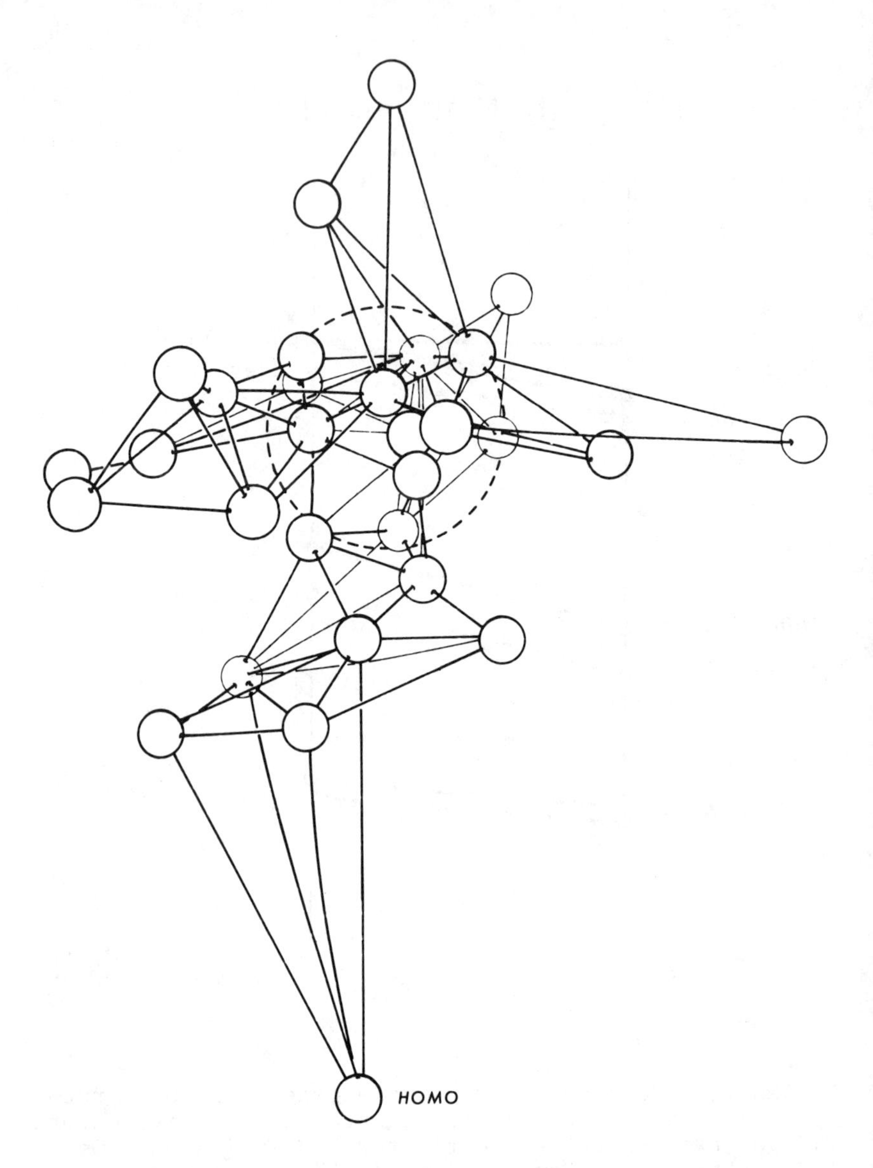

FIGURE 15: The three-dimensional model of lower limb proportions shown in figure 11 after rotation through ninety degrees demonstrates that humans lie uniquely separate from all of the non-human primates. This is also the case for other studies of elements of the lower limbs.

Overall Bodily Proportions

It is, of course, possible to look at other parts of the body using these techniques. Such studies are described elsewhere (Oxnard, 1981a, 1982). However, with the aforementioned results in mind, it is obvious that a key question is: what happens if we meld all of the anatomical regions? For technical reasons this cannot yet be done with the regional studies of limb girdles, limb segments, cheiridia and so on; it can, however, be carried out with the overall proportions taken by Professor Schultz on upper and lower limbs, trunk, and head and neck. The result of multivariate statistical analysis of his data is summarized in the minimum spanning tree of figure 16.

First, it is apparent that the functional arrangements of the regional studies have now been superceded. Second, it is immediately obvious that the pattern of genera seen in the first three figures has returned to us. The genera are arranged in a more or less linear sequence. Prosimians are at one extreme, next lie New World monkeys, next monkeys of the Old World, next apes, and, at the opposite extreme to prosimians, humans (figure 16). These taxonomic groups are most evident though, within each, further studies show subgroupings that are a little more controversial. Within New World monkeys (figure 17), for instance, the subfamilies of the present day are clearly recognizable (although it is true that there is a superclustering that is more controversial to classical taxonomists). Within Old World monkeys (figure 18) supergeneric groupings that make taxonomic sense are also clearly delineated (though, again, total agreement with classical taxonomy is not found). Even among the hominoids lesser apes are separated from great apes (although, once more, classical studies are not entirely followed as the separations within the larger bodied hominoids are, figure 19, between the orangutan and the gibbons on the one hand, and the African apes on the other).

It is only among the prosimians that marked taxonomic tangles appear to occur. For although the lorisines and lemurs seem reasonably associated, tarsiers appear closely linked with bush-babies and aye-ayes appear <u>not</u> closely linked with indriids - two features of the results which do not agree with current ideas about the relationships of these forms.

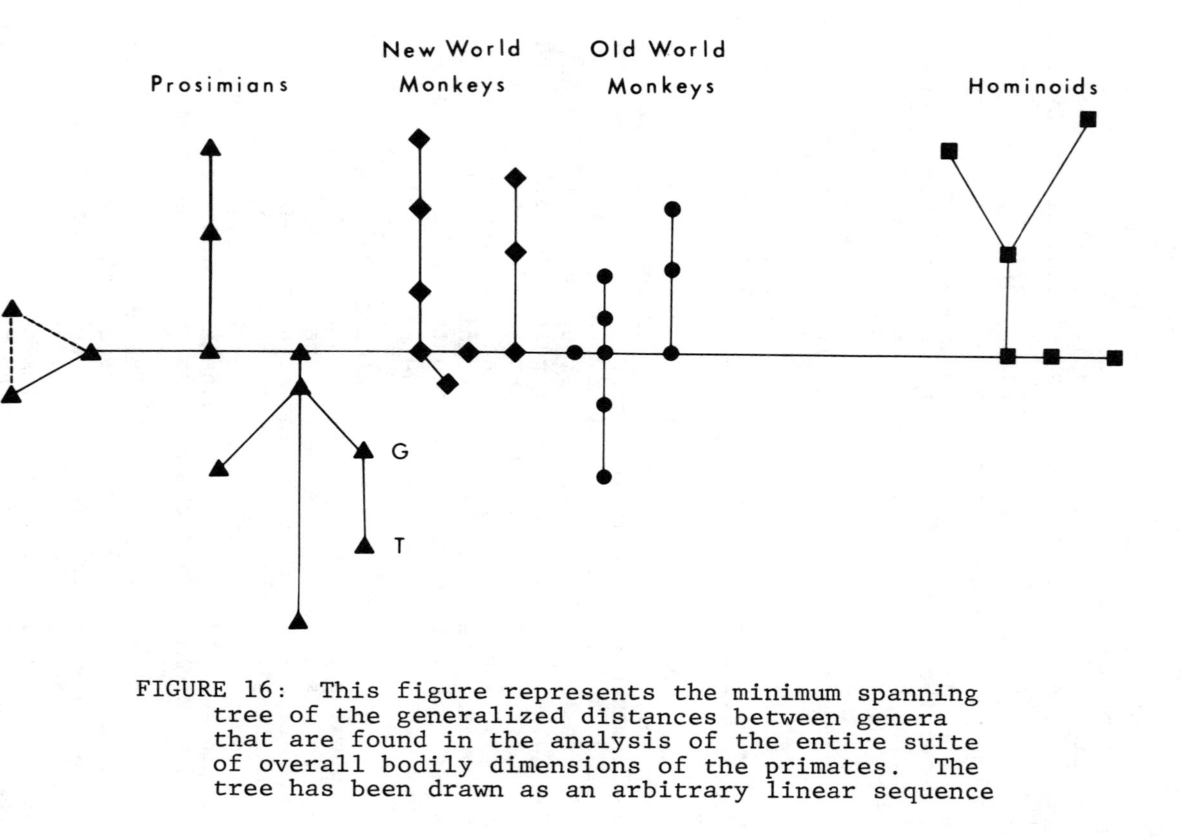

FIGURE 16: This figure represents the minimum spanning tree of the generalized distances between genera that are found in the analysis of the entire suite of overall bodily dimensions of the primates. The tree has been drawn as an arbitrary linear sequence

Legend continued:

traversing from one end to the other. Single branches off the main skeleton have been drawn at right angles to it, secondary branches, by convention, at angles of sixty degrees. The diagram enables us to see that not only does this complex result readily identify the major taxonomic groups in linear sequence (as per figures 1, 2 and 3) but so also does it identify, relatively correctly, many of the taxonomic subgroups of the major divisions of the primates. Thus squares (all hominoids) are linked together, circles (Old World monkeys) are so linked, as are all diamonds (New World monkeys) and all triangles (prosimians).

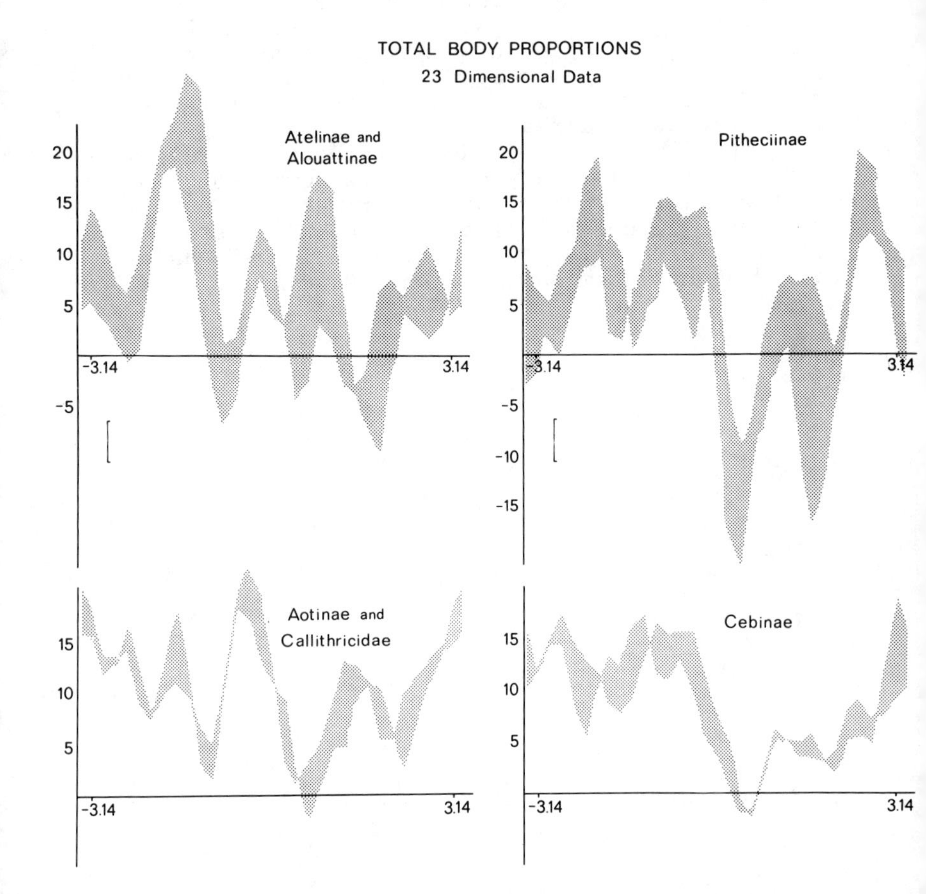

FIGURE 17: This figure utilizes Andrews' high-dimensional method to display the total pattern of canonical axes for the study of overall bodily proportions in the New World monkeys. In contrast to studies of the limbs alone (figures 7 and 12) the result does not seem related to locomotor function. Like the study of the entire body for the entire primates (figure 16), it makes sense when seen in the light of current biomolecular relationships (Oxnard, 1981a).

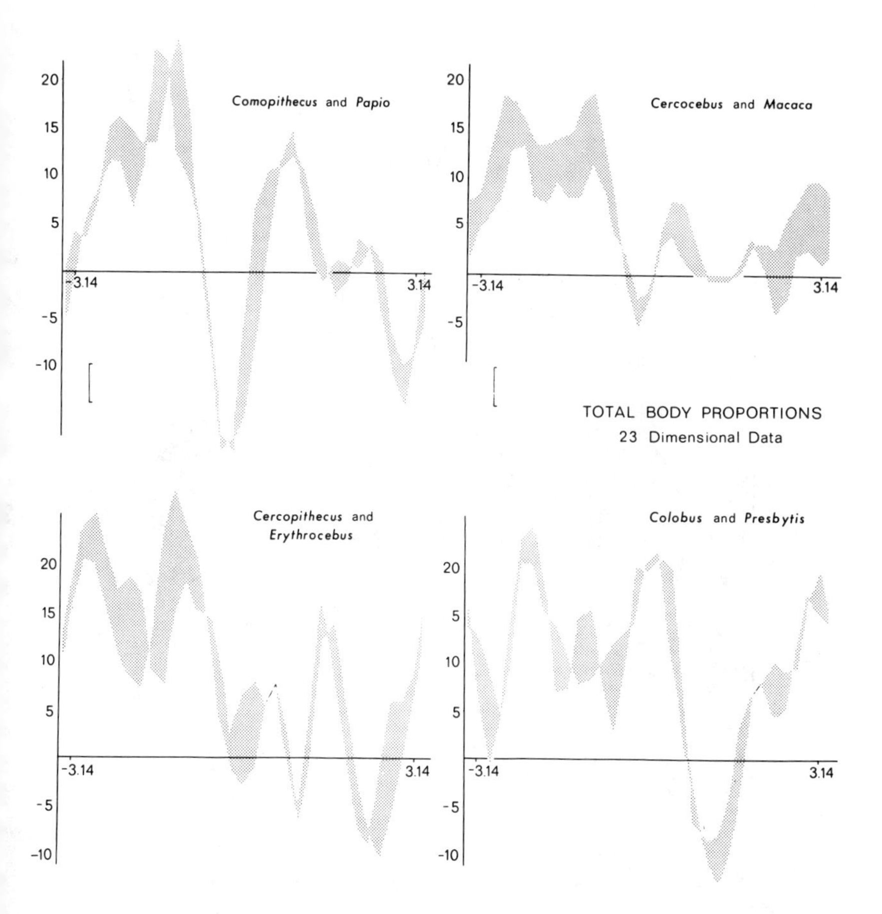

FIGURE 18: This figure takes the display of figure 17 and applies it to the study of overall proportions for the Old World monkeys. Again, in contrast to studies of limbs alone (figures 7 and 12), the relationships here do not seem to parallel locomotor function. Like the study of the entire body proportions for the entire primates, they parallel most closely the biomolecular relationships of the genera (Oxnard, 1981a).

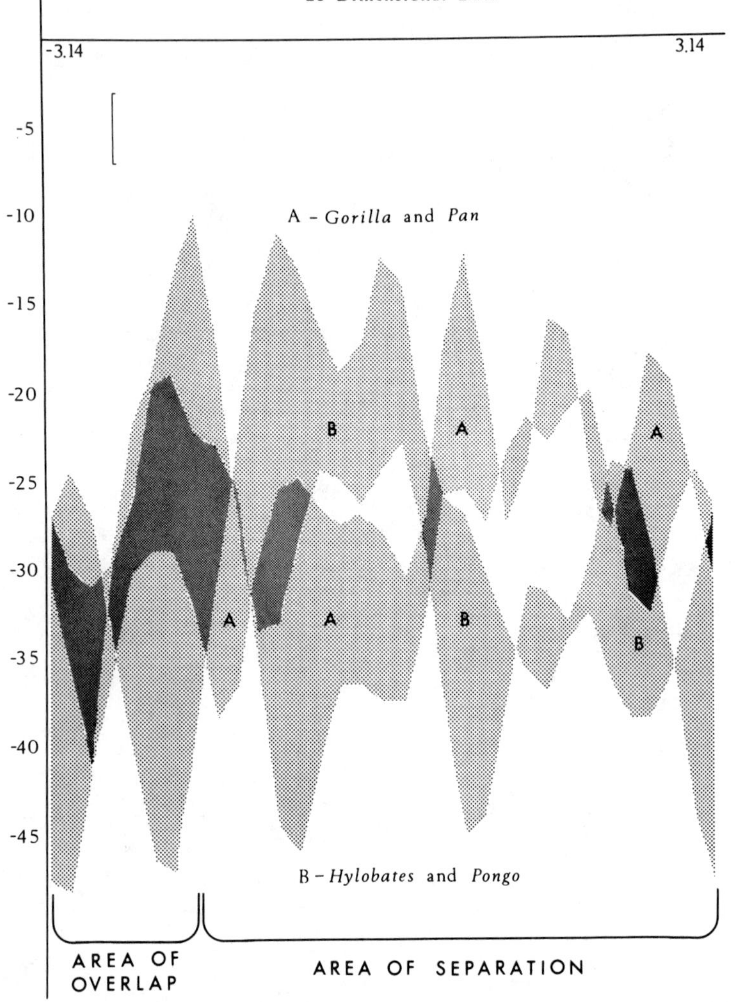

FIGURE 19: This figure utilizes once again Andrews' high-dimensional display and applies it to the total overall proportions of the apes. The result does not parallel function such as is found in the study of limbs and limb parts. It does follow taxonomy such as is found in the study of overall

Legend continued:

proportions of the entire primates, although once again, the taxonomy that it seems to copy is not classical (which would ally lesser apes against great apes against humans), but molecular (allying African apes [and humans] against orangutans together with lesser apes, Oxnard, 1981a).

Let us first look at the _Tarsius-Galago_ link. These are two genera that are not only not within the same family or super family, they are, according to many, not even within the same infraorder. And at least a few authors (but an increasing number nowadays) would place the tarsiers with the monkeys, apes and humans in the anthropoidea (or haplorrhini).

However, if we scrutinize the information about tarsiers more closely, another feature of the data obtrudes itself. That is, though the link between tarsiers and bush-babies as figured here is indeed the minimum link, the next closest links tell a different story. The next nearest links of bush-babies are indeed with a series of prosimians (especially mouse lemurs). But the next closest links of tarsiers are with a variety of anthropoids, most prominently some New World monkeys: owl monkeys, tamarins and capuchins, and almost equally distantly, some Old World monkeys: vervets and mangabeys (figure 20). This finding should be set along side those of Cave (1973), Groves (1974), Minkoff (1974), Szalay 1975a and b), and Luckett (1975). It seems at least possible that the existence of the minimal link between bush-babies and tarsiers is because the functional convergence between them overshadows all else. But the existence of totally different other neighbors for each may be speaking most cogently to what remains when functional convergence is dissected away (Oxnard 1978b). That remnant could well relate to the true evolutionary affinities of tarsiers.

Let us, second, look at the non-alignment, in our results, of aye-ayes with indriids. The conventional view is that the extant species of _Daubentonia_ (together with its closely related sub-fossil species, _D. robustus_), is closely allied with lemurs, especially the indriids. But new doubts have been expressed by Groves (1974) and by Jouffroy (1975) on the basis of a variety of individual features from many regions of the body. Most recently of all, Oxnard (1981b) has assembled a very large number of anatomical dimensions (not characters) of all regions of the body, and examined them using multivariate statistical analysis. The subsequent separations of _Daubentonia_ from all other primates in every and all combinations of features suggests that it is truly unlikely that the genus is close to regular prosimians. If this point, also, is further confirmed, then even this apparent conflict with ideas about the relationships of the primates is removed.

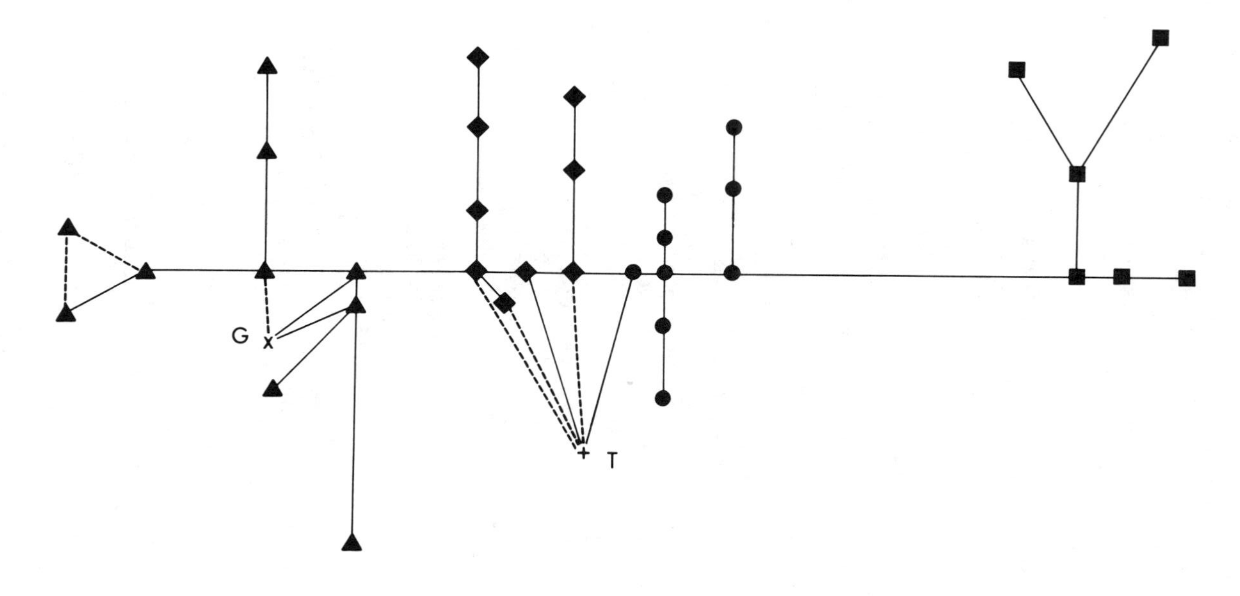

FIGURE 20: This shows a view of figure 16 in which the link between _Tarsius_ and _Galago_ has been removed. The effect of this is to move _Galago_ (X) to a new position clearly embedded among prosimians (triangles), and to move _Tarsius_ (+) to a new position close to some New World monkeys (diamonds) and some Old World monkeys (circles.)

It is, thus, remarkable that these multivariate studies of primate measurements should mirror the general arrangements of the primates that are evident in the classical morphological, physiological and biomolecular approaches to the problems of primate evolution outlined in the introduction. What could be the possible reasons?

One may be circular. Perhaps in some non-quantitative way systematists have already included in their investigations information about the overall proportions of the body. Careful study of the classical literature on the primates reveals the exact opposite. Judgements have not been made on the basis of the overall dimensions of the animals whether by ruler or eye. In fact a rather overt attempt has been made to avoid such measures exactly because they have been thought to be too closely linked with function. The morphologies that have been used are truly those outlined in the introduction: that is, features of the skull (teeth, jaws, orbits, middle ears, cranial bases, and so on) and of the external form of the animals (faces, cheiridia, genitalia, pelages and so on). Those previous examinations of overall bodily form that have been carried out (e.g. Mollison, 1910, Priemel, 1937, Schultz' earlier work summarized in 1969, Erikson, 1963) provide functional results, not systematic ones. It is the multivariate comparison of these data, including patterns of variance and covariance among the proportions, that gives the new picture.

Presumably, therefore, the correspondences obtaining between these multivariate statistical results of analysis of overall bodily proportions and those stemming from morphological, physiological and biomolecular investigations must stem from the notion that, whether or not the individual parts reflect function, the addition of many such parts provides information resulting in evolutionary relationships.

Even this is not the whole story. In those parts of the results where classical organismal studies conflict with biomolecular investigations, it is with these latter that the multivariate morphometric approach is most congruent (Oxnard, 1981a). This has further implications for understanding the relationships of the primates (Oxnard, 1982).

Whatever the ultimate rationale for these results, however, the findings provide both marked encouragements and dire warnings for the study of primate fossils. They give encouragements because, whenever it is possible to add together, using multivariate statistical methods, many different pieces of information, we may well hope for the evolutionary result. They provide warnings because they make it abundantly clear that study of individual anatomical regions of fossils is unlikely to give much information about fossil evolution and taxonomy. A fragment of clavicle alone, the lower end of the radius in isolation, a single pelvis, talus, or toe bone tell us primarily about the functions of clavicles, radii, pelves, tali or toe bones, respectively, within a wide and unknown range of behavioral possibilities for the fossil.

Implications For Australopithecines

What then can be the implications of these various studies for the assessments of a series of finds like those of *Australopithecus africanus* (synonyms are *Homo habilis*, *H. africanus*) as at Olduvai and Sterkfontein, and of *A. robustus* (synonym, *Paranthropus*) as at Kromdraai and Makapansgat?

Our prior studies (reviewed in Oxnard, 1975a,c) have shown that these australopithecines (but not necessarily the finds of more recent years, such as from the Afar valley that have also been attributed to the same taxa) are quite unlike the equivalent parts of humans.

Studies of the primate pelvis demonstrate that the australopithecine pelvis is quite disimilar to that of humans and, although very possibly capable of extended bipedal movement, could scarcely have been bipedal in human manner; relationships between joints are somewhat human, muscular leverages and orientations are ape-like (Zuckerman, Ashton, Flinn, Oxnard and Spence, 1973; Ashton, Flinn, Moore, Oxnard and Spence, 1981).

Investigations of australopithecine tali suggest that even if these were possessed by animals capable of a degree of bipedality (Day and Wood, 1968), it is rather likely that the animals were also capable of a degree of climbing activity in a way totally denied to humans today (Lisowski, Albrecht and Oxnard, 1974, 1976; Oxnard 1980).

Study of the entire Olduvai foot using sections and reconstructions confirms this view (Lisowski and Oxnard, 1980; Oxnard, 1980); this is independently corroborated by the comparative arthrological approach of Lewis summarized in his 1981 review.

Study of fragments of an australopithecine clavicle and scapula suggests that they, too, have some of the hall marks of a creature able to bear tensile forces in its upper limb (summarized in Oxnard, 1975a,c). The clavicle possesses the longitudinal torsion characteristic of many animals that hang by their upper limbs (and this torsion is found throughout the length of the bone so that it is evident even though the ends of the bone are missing in the fossil). The scapula has a set of the glenoid fossa upon the axillary border (again, something that can be determined, when specially defined as in these studies, even though the entire axillary border is not present) that is likewise characteristic of species that regularly bear tensile stresses in the shoulder region.

Examination of other anatomical regions (toe bones, metacarpals, perhaps phalanges, certainly humeri, summarized in Oxnard 1975a,c) suggest that their owners were uniquely different functionally from both the humans and African apes of today. These differences are so great that the australopithecines are, in these features, actually more different from humans and African apes than these latter are from one another.

What do we mean by saying that the australopithecines are uniquely different from both living humans and living African apes? For it is obvious that in one sense every type of primate is uniquely different from every other. We are all sensitized to the idea that each human being is uniquely different from every other. Certainly, gorillas may be said to be uniquely different from chimpanzees. No one could possibly mistake, for instance, gorilla structures for chimpanzee structures. But gorillas and chimpanzees, in comparison with other hominoids, are not uniquely different from each other; gorillas, in many ways, are chimpanzees writ somewhat larger, although this is not, of course, the only difference. When we have described the structure of gorillas we have also described much of the structure of chimpanzees. Likewise, we would also expect that near ancestors of each, although different from each, would also be similar to both in comparison with differences from other hominoids.

In the same way, reasonably close ancestors of humans, though not yet human, should also possess similarities with humans in comparison with other hominoids, although they would not, of course, be identical to modern humans. In the multivariate statistical studies of individual australopithecine fragments, the australopithecines do not fit that description; they are more different from both apes and humans than these latter are from each other.

And it is the addition of some features of the pelvis in which australopithecines resemble humans to other features in which they resemble apes that gives a picture of the entire pelvis in which it is uniquely different from either humans or apes. A similar view arises from studying the various other parts of the post-cranial anatomy of the australopithecines using such methods. (Oxnard, 1975c,1979c).

The interpretation of such differences is not easy. If the australopithecines were truly intermediate in structure, we could reasonably posit a degree of intermediacy in function; and we might expect this in pre-human ancestors. But, given that, in the post-cranium, these fossils are unique, we can only suggest functional uniqueness.

Functional uniqueness could mean a totally unique and therefore unknown manner of locomotion and posture; we may judge that as unlikely.

But their funtional uniqueness could also mean something else. Thus, just as their structural uniqueness depends upon a set of structural features that are only unique when taken in combination, so their functional uniqueness might depend only upon a curious combination of functional attributes. Thus it is entirely possible that the australopithecines possessed a mode of locomotion and posture no one element of which was unique among the hominoids, but the combination of all in a single species was.

There is some evidence from the pelvis and foot that they could move bipedally. There is also very clear evidence from the pelvis and foot that they could move quadrupedally. Evidence from the upper end of the femur is especially strong in this regard. The upper limb, as evidenced in the shoulder and hand, for instance, may well have been able to bear tensile forces suggesting acrobatic types of climbing activity;

the re-articulated foot suggests the same. Some extant non-human hominoids can move very well quadrupedally, other extant non-human hominoids are especially good at climbing in an acrobatic manner, only humans can walk bipedally in an established fashion. None of these individual patterns is unique to hominoids; but their combination in a single hominoid certainly would be. This is the direction in which the combination of features in the Olduvai and Sterkfontein australopithecines seems to be pointing.

Even the new paleoecological evidence now allows such an interpretation. For although it is recognized that the times during which the australopithecines lived were basically semi-arid, the generally accepted interpretation - that there were no trees (and therefore no climbing environments) - is probably wrong. New paleoecological evidence (for example, Butzer, 1976) and, indeed, biological common sense suggests that, even in relatively dry times, every possible type of environment exists within a few miles of many rivers: from immediate riverside wooded areas, through heavy but entirely local forest, fading to thinner woods, through isolated forest stands, through single large trees, through smaller trees and bushes, to the relatively treeless savannah. We cannot tell in which part of a similar habitat the australopithecines lived. Indeed, they may have been adapted for all.

The series of studies described above throws considerable doubt upon the conventional characteristics of these australopithecines as human ancestors although we must remember that the conventional wisdom is yet that _Australopithecus_ is on the main human lineage. This view is particularly enunciated in the historical review provided by Professor Charles Reed in the introduction to this volume. It is not denied, apparently, by a small number of unrelated investigations that seem to point in an opposite direction (e.g. the non-human form of [a] the australopithecine ear ossicles, Rak and Clark, 1979, and [b] the odontometrics of the dentition, Hansinger, 1976).

What criticisms may there be of these new ideas? For although they have been produced by many complex and difficult studies over several decades and involved much comparison with large numbers of extant species, they may well still be thought too limited in scope to convert most paleoanthropologists. The fossil record is apparently now so complete and well-studied, we are told, that multivariate studies of less than a dozen isolated fossil finds are unlikely to change many minds. Several comments result from this notion.

Although there have indeed been a great many studies published in the years since their first discovery, except for the new finds of the last half dozen years, the entire output of the twentieth century fossil hunters has not provided any very large number of useful post-cranial fragments for study. It has only been in the last decade that really large finds have been made (e.g. Richard Leakey at East Turkana, Mary Leakey at Laetoli, Johanson and Taieb in Ethiopia, Clark Howell in the Omo).

And many of these new finds seem to be things other than australopithecines like those of Olduvai and Sterkfontein. Some are, for instance, very much older than these australopithecines; and some seem to be very much more like humans than traditional australopithecines. Thus the skull (ER 1470) found at East Turkana has a much bigger cranial capacity (approximately 800 ccs) compared with australopithecine skulls almost a million years younger at 450 ccs (Leakey, 1973). A talus, also from East Turkana, older by nearly two million years than the Olduvai talus, is far more like human tali than are the Olduvai and Kromdraai specimens (Wood, 1973). A footprint more than two million years older than the Olduvai foot is very human in its form (Day and Wickens, 1980) and unlikely to have been made by a foot resembling the Olduvai foot (Oxnard, 1980). A humeral fragment from Kanapoi may be four or five million years old, yet is almost indistinguishable in shape from many individual human humeral fragments (Patterson and Howells, 1967; McHenry, 1973; Oxnard 1975a; Senut, 1981); and this compares with australopithecine humeral fragments of little more than a million years old that are completely different from modern humans (Oxnard, 1975a; Senut, 1981).

Most of the new fossils (e.g. those found in Ethiopia, Johanson and Taieb, 1979) have not yet been fully studied. Though they have been called "australopithecines" (A. afarensis) we must be aware that they do not seem to be similar to the later australopithecines of Sterkfontein and Olduvai. Indeed, in so far as the new studies turn out to be correct, they emphasize my view that the australopithecines discussed in this chapter are not on the human lineage. Johanson and White (1979) believe, for instance, that they represent a new species ancestral to both later australopithecines and to humans.

However, it is almost as equally likely that this claim about A. afarensis will turn out to be false because, once again, the discoverers of a particular set of fossils have been unable to resist the idea that their finds are at the source of the human lineage. When we remember that this same claim has been made each time something has been discovered (for Piltdown man, for example, and how wrong that turned out to be, for the Taungs child, now discredited as at the beginning of the lineage, for the elder Leakey's Olduvai finds, now apparently overturned by the Afar specimens, and so on) we are likely to remain sceptical until shown otherwise by independent research and the light of future finds. Perhaps one of the first of these independent investigations may be the report by Senut (1981) that the afarensis humerus at approximately 3 million years has a pongid like structure; it contrasts markedly with the Kanapoi specimen at 4 million years that she, too, shows is far more human-like.

That we should remain sceptical is particularly evident from new finds in China. Again, though these have been by no means fully studied, Woo (1981) reports on extensive coverage in Chinese paleoanthropology of the entire spectrum of human and pre-human fossils. Thus, in addition to both early and late Homo sapiens in many sites (e.g. Guangxi and Hubei provinces) Homo erectus has been widely found (e.g. Anhui, Hubei and Shaanxi provinces, and especially at 1.7 million years at Yuan Mou in Hunan, and therefore far older than many of our australopithecines). Teeth resembling australopithecine teeth are now known from Hubei and Shaanxi provinces. Preliminary reports suggest that 8 million year old Ramapithecus and Sivapithecus have been located in Yunnan. Even Dryopithecus (a few teeth from Yunnan) may exist in China. All of this supplies, if eventually confirmed, information that supports the non-human view of the australopithecines of Sterkfontein and Olduvai. It even brings into question the matter of Africa as the site of human genesis.

However, my observation: about multiplicity of studies yet paucity of materials, does not apply to these new finds that have not yet been fully studied. It applies to the morass of older investigations of crania, jaws and teeth on which the conventional picture depends. It is necessary, therefore, to go back to those earlier studies to see if, in truth, they totally negate the new assessments.

There are good reasons for doing this. For the earlier controversies drove some of the earlier participants into untenable positions. In the uproar, at the time, as to whether or not these creatures, australopithecines, were truly like humans, the opinion, that they are, won the day. This resulted not only in the defeat of the contrary opinion, but also in the burying of that part of the data on which the contrary opinion was based. It should be possible to unearth this other part of the evidence.

The various controversies over the shapes and sizes of teeth are a case in point. Although many studies were carried out, the general consensus is that the evidence from the teeth clearly supports the human-like status of these fossils. What, however, are the facts? Study of the work of Ashton and Zuckerman (1950) and of Bronowski and Long (1952) followed by Ashton, Healy and Lipton (1957) shows that in some features (incisors, canines and first lower premolars) the australopithecines are human-like (and this is well publicized). But these studies also demonstrate that in other features (remaining premolars and molars) the fossils are ape-like (and this has generally been disregarded). Investigation, with hindsight, of one of the results (for the milk canine) shows it to be neither human-like nor ape-like, but uniquely different from each (figure 21). In the light of the foregoing discussion of mixtures of features in the post-cranium, these older results are not, in fact, evidence of human-like status (or, for that matter, ape-like status), but are most compatible with the idea that the creatures are unique.

Another example of such controversy stems from various studies of the basicranium. The fact that the spheno-ethmoidal angle of the australopithecines is markedly similar to that of humans has been seized upon in many prior arguments (e.g. Tobias, 1967). The fact that the foramino-basal angle of the fossils is markedly similar to those of apes has been neglected (table 1). This is another example where the reality is a combination of features, some like humans and others like apes, in toto therefore, unique. This is confirmed by a recent restudy of this area (Ashton,

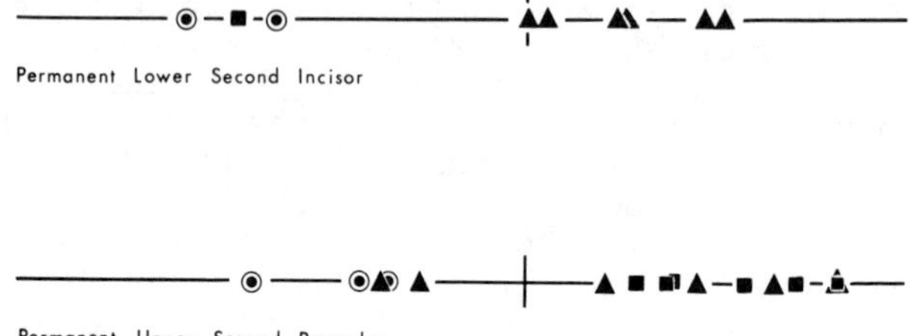

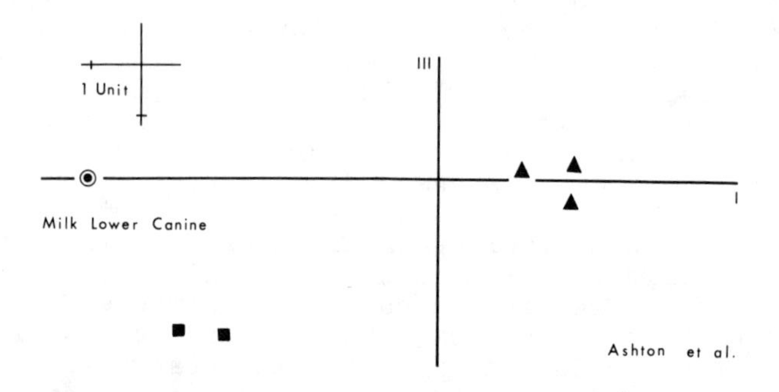

FIGURE 21: A representation of the results of Ashton, Healy and Lipton, (1957). Similarities of the fossils (squares) with humans (circles) is evident in the multivariate statistical analysis of the dimensions of the incisor tooth. Similarities of the fossils with apes (triangles) occur in the studies of premolar and molar teeth. In the dimensions of the milk lower canine, the fossil is uniquely different from both humans and apes as shown by its separation in an additional multivariate statistical axis. The standard deviation marker applies to all studies.

TABLE 1

HOMINOID BASICRANIAL AXIS

Regression Adjusted

Genus	Anterior Limb	Posterior Limb	Spheno-Ethmoidal Angle	Foramino-Basal Angle
Homo	57	33	**133**	131
Gorilla	60	27	172	**121**
Pan	57	29	156	**121**
Pongo	53	31	165	**116**
A. africanus (Sts 5)	51	29	**129**	**109**
A. robustus *(Zinjanthropus)*	64	25	121	128

(S.E. always <2% of mean)

Flinn and Moore, 1975) that shows that, in multivariate combination, the fossils are indeed unique.

Yet another example includes the facial area and its contained infraorbital foramina and nerves (Ashton and Oxnard, 1958). The fact that there is only a single foramen in the fossils, and that in this feature they resemble humans, has been recorded many times, (e.g. Clark, 1959) since the early mention by Broom and Schepers (1946). The other half of the evidence is that in the position of the foramina upon the face the fossils are much more like the apes (Ashton and Zuckerman, 1958). Taking these two sets of contrasting features together proscribes a unique situation in the fossils (table 2).

And yet another example still that may be cited includes a new study of the condylar nuchal region of the skull (Adams and Moore, 1975). This shows that the position of the occipital condyles in the fossils is convincingly ape-like. But this feature is combined with an angulation of the condyles upon the skull base that is remarkably human. These facts may possibly relate to upright posture in a creature that is also a capable quadruped or climber. Once again, however, the new view, that the creature is unique, is supported (table 3).

One final example is the question of the form of the articular eminence in humans, apes and fossils. The original emphasis was placed upon the supposed human-like position of the articular eminence in the fossils (e.g. Clark, 1959). But other studies emphasized the ape-like proportions of this feature in the fossils (Ashton and Zuckerman, 1954). Once again, this curious combination of features renders the australopithecines uniquely different from both humans and fossils as is shown in a multivariate study by Ashton, Flinn and Moore (1976).

It seems clear that there should now be carried out a whole series of new investigations in which, first, all aspects of the data should be included and in which, second, data should be assimilated in some overall form (such as by a multivariate statistic) so that the effects of combinations of features can be understood.

It is necessary to look for other ways of displaying these results. For many of the ideas carry less weight in the minds of some scholars because they are expressed through this relatively new technology.

TABLE 2

INFRAORBITAL FORAMINA

Genus	Number of Foramina	Position of Foramina[1]
Homo	1·05	21
Gorilla	1·60	45
Pan	2·40	40
Pongo	3·30	35
A. africanus	1·00 (in 8)	36 (Avg. of 2)
A. robustus	1·00 (in 4)	59 (Avg. of 3)

[1]Percent distance from orbital to alveolar border.

TABLE 3

CONDYLAR-NUCHAL REGION

	Condylar Position Index	Condylar Angulation	Nuchal Insertion Index
Homo	74	**86**	106
Gorilla	**27**	62	39
Pan	**28**	65	44
A. africanus (Sts 5)	**37**	**85**	67
A. robustus (*Zinjanthropus*)	50	92	62

(S.E. always <2% of mean)

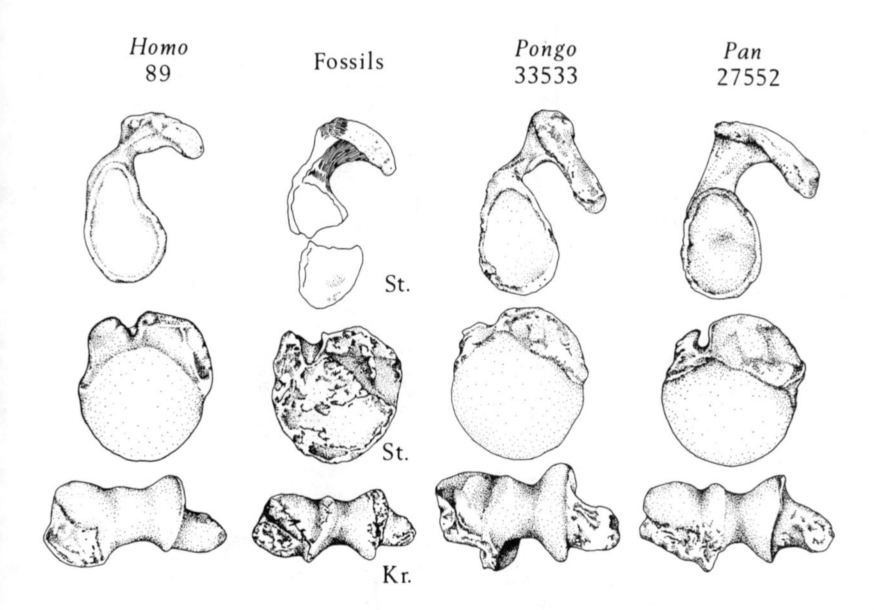

FIGURE 22: Articular surfaces of upper limb elements for three particular specimens of Homo, Pan, and Pongo. The specimens were chosen on the basis of having similar glenoid dimensions as the fourth specimen: a fossil fragment of Australopithecus from Sterkfontein. On this basis, the other articular upper limb surfaces shown (upper and lower ends of humerus) are of similar proportions among all the extant forms. It is perhaps merely fortunate that the articular elements for upper and lower ends of the humerus for the fossil are also of approximately similar size, because, of course, these do not come from the same fossil individual. The diagram shows that, given a similar scaling for one upper limb articular surface, similar scalings for other articular surfaces exist among the extant species. And it shows that the fossil articular surfaces, though from different individuals, also fit this picture of similar relative proportions.

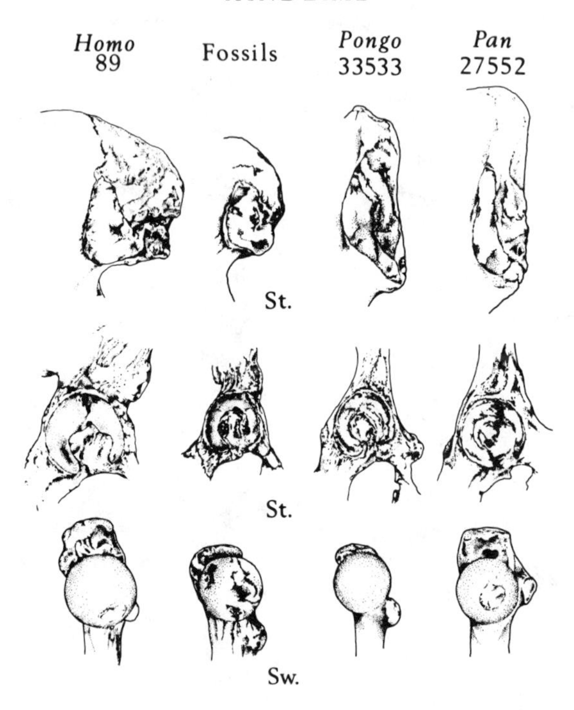

FIGURE 23: Diagram of the pelvic and femoral lower limb articular surfaces for the same specimens of Homo, Pan and Pongo shown in figure 21 and based upon the same set of dimensions (i.e. scaling the specimens so that the glenoid fossa dimensions are similar). This diagram demonstrates that, in comparison to upper limb articular surfaces, human hip articular surfaces are larger than those of apes. This presumably relates to lower limbs in humans bearing all of the body weight, whereas in apes that body weight is shared with more equally sized upper limb articular surfaces.

Is it an accident only that each of the fossils specimens resembles proportions in apes rather than in man (the fossil specimens are not, of course, from the same specimens)? It could be chance alone that has provided us with upper limb articulations from big specimens of the fossils and fossil hip articulations from small specimens of the fossils.

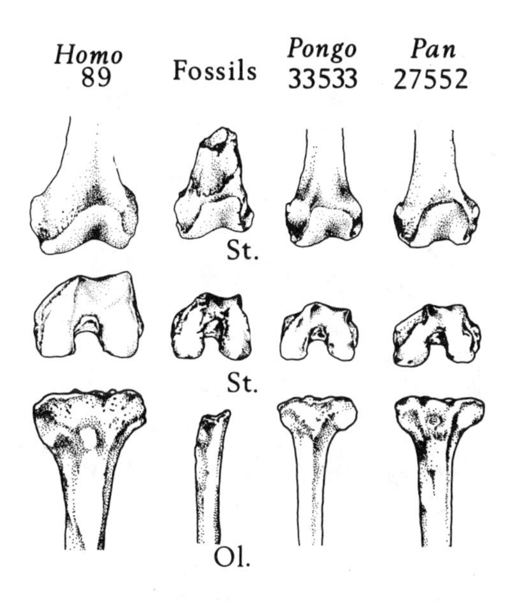

FIGURE 24: Articular surfaces from around the knee for the same specimens of Homo, Pan and Pongo shown in figure 21 and based upon the same set of dimensions (i.e. scaling the specimens so that the glenoid fossa dimensions are similar). This diagram demonstrates that, in comparison to upper limb articular surfaces, human knee articular surfaces are also larger than those of apes. This is similar to the finding for hips and presumably relates to lower limbs in humans bearing all of the body weight, whereas in apes that body weight is shared with more equally sized upper limb articular surfaces.

Is it an accident only that each of the fossils specimens resembles proportions in apes rather than in man (the fossil specimens are not, of course, from the same specimens)? It could, I suppose, still be chance that has provided us with upper limb articulations from big specimens of the fossils and both hip and knee articulations from small specimens of the fossils!

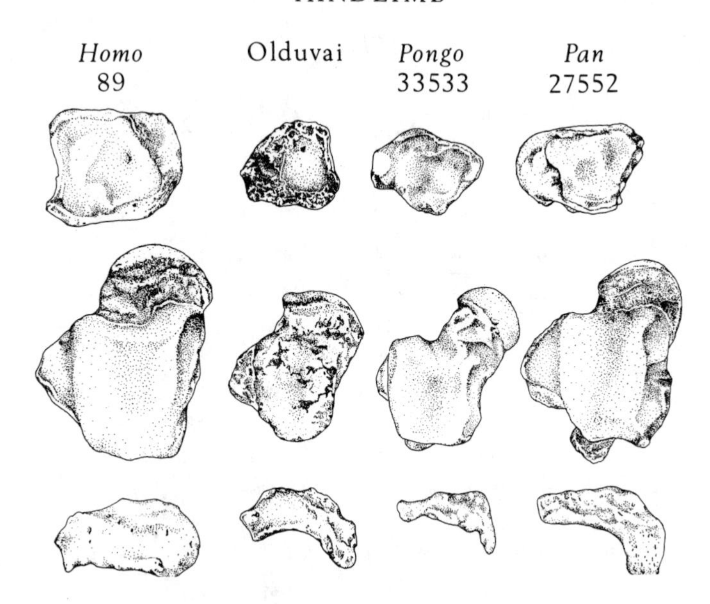

FIGURE 25: Articular surfaces from the ankle and foot for the same specimens of Homo, Pan and Pongo shown in figures 21 and 22, and based upon the same set of dimensions (i.e. scaling the specimens so that the glenoid fossa dimensions are similar). This diagram demonstrates that, in comparison to upper limb articular surfaces, human ankle and foot articular surfaces are larger than those of apes. This is similar to the findings for both hip and knee, and presumably relates once again to lower limbs in humans bearing all of the body weight, whereas in apes that body weight is shared with more equally sized upper limb articular surfaces.

Surely chance cannot now be responsible for the fact that the fossil articular surfaces of hip, knee, ankle and foot all display relative proportions that ally them with the apes rather than humans! Surely the combination of chances involved in figures 23, 24 and 25 are beyond coincidence! The differences are so great that measurements are scarcely needed to demonstrate them.

(multivariate statistical analysis) - a technology that is not easily understood, is easily mishandled, and is not readily accessible to those who do not themselves utilize it (Oxnard, 1978c). An excellent exposition of its use has been given recently by Albrecht, (1980).

For this reason I have attempted to provide appropriate visual comparisons of some of the individual bones (Oxnard, 1975a, 1977b). These seem to confirm the non-human, ape-like status of the australopithecine post-cranial fragments in a new way. Thus, though classical studies of the articular surfaces of long bones of australopithecines have been interpreted as showing them to be human-like in character (e.g. as reviewed by Robinson, 1972), a simple visual comparison (figures 22 through 25) of differences in proportion between upper and lower limb articular surfaces is in marked contrast.

These figures show most convincingly that human lower limb articular surfaces are large in comparison with human upper limb articular surfaces; this presumably befits their bipedal status where the lower limb takes all of the body weight in locomotion.

These figures also reveal that ape upper and lower limbs have a more equal relationship in proportions and this presumably relates to the more equal sharing of the body weight in the more frequent four-limbed habits of these creatures, whether on the ground in knuckle walking, or in the trees during climbing.

It is of extreme interest that, in such a set of comparisons, the information that is available for the fossils shows them _not_ to have the large lower limb articular surfaces of humans. They have more equal upper and lower limb articular proportions, and thus resemble more the apes. This should be set alongside the comment of Richard Leakey who reported (in 1971) that "preliminary indications point to a relatively short lower limb and a longer forelimb" for _Australopithecus_. One can only come to the conclusion that, however able these creatures may have been on two legs, they were also convincingly able to move on four; and this is a feat denied to humans today (their ontogenetic quadrupedal state is inefficient in the extreme).

Broader implications for human evolution

There is thus now a rather strong body of information stemming from the combination of skeletal features that relegates the conventional australopithecines of Olduvai and Sterkfontein to a side role in human evolution.

This raises a series of further possibilities.

First, it is rather likely that Homo as a genus has existed for an order of magnitude longer than previously thought even only ten years ago (five million years, probably even much more, rather than a mere half million or so). Homo must surely have been contemporaneous with many of the australopithecines. Such a new time span for the genus is so much greater than that generally believed to be the case that very much more time may have been available for the psycho-social evolution of the genus. It is possible, therefore that those complex features of humans in which they differ most markedly from animals: behavioral, cultural, intellectual, creative, may depend rather less upon the prior genetic status of animal forebears, and very much more upon their own psycho-social evolution during this extra amount of psycho-social evolutionary time.

Second, the simpler abilities of humans that relate to bipedality and a free upper limb may also have existed for far longer than we have previously thought. That in turn almost certainly means that these structural features may have evolved not once, not twice, but perhaps several times. Such a set of evolutionary parallels or of radiations would closely resemble what zoologists recognize as sudden evolution into new adaptative zones or spread into new niches (e.g. Simpson, 1953; Gould and Eldridge, 1977). It would also inevitably mean however, that the genus Homo as we know it at the present time is the only remaining evolutionary experiment along bipedal lines; other genera have been unsuccessful.

Finally, these ideas may also mean that the evolution of the various australopithecines may be a most interesting separate set of problems in their own right. What were the actual behaviors of the various creatures that make up this particular group of species? What were their relationships with humans, and with apes? We have attempted some answers. They presumably represent some of the essays into bipedalism that failed. They presumably are neither on ape nor human evolutionary lineages, but a third group in their own right. We may be able to see more clearly the flowering (and withering) of this entire group of creatures once we are able to remove from our minds the idea of australopithecines as direct human ancestors.

Acknowledgments

The ideas expressed in this paper have come from discussion of results of many studies with colleagues, particularly Professor E. H. Ashton, University of Birmingham, and Professors W. J. Moore, University of Leeds, and F. P. Lisowski, University of Hong Kong. Most of all, however, they stem from the initial stimulus and ever continuing interest of my own major Professor, Lord Zuckerman, OM, KCB, MD, DSc, President of the Zoological Society of London and Honorary Professor, University of East Anglia, U.K.

References

Adams, L. M. and Moore, W. J. 1975. Biomechanical appraisal of some skeletal features associated with head balance and posture in the hominoidea. Acta Anat. 92:580-594.

Albrecht, G. H. 1980. Multivariate analysis and the study of form, with special reference to canonical variate analysis. Amer. Zool. 20:679-693.

Andrews, D. F. 1972. Plots of high dimensional data. Biometrics 28:125-136.

Andrews, D. F. 1973. Graphical techniques for high dimensional data. In: Discriminant Analysis and Application. Ed. T. Cacoullos. Academic Press, New York, pp. 37-59.

Ashton, E. H., Flinn, R. M. and Moore, J. W. 1975. The basicranial axis in certain fossil hominoids. J. Zool. Lond. 176:577-591.

Ashton, E. H., Flinn, R. M. and Moore, J. W. 1976. The articular surface of the temporal bone in certain fossil hominoids. J. Zool. Lond. 179: 561-578.

Ashton, E. H., R. M. Flinn, J. W. Moore, C. E. Oxnard, and T. F. Spence. 1981. Further quantitative studies of form and function in the primate pelvis with special reference to Australopithecus. Trans. Zool. Soc. Lond. 36:1-98.

Ashton, E. H., R. M. Flinn, and C. E. Oxnard. 1975. The taxonomic and functional significance of overall body proportions in primates. J. Zool. Lond. 175:73-105.

Ashton, E. H., R. M. Flinn, C. E. Oxnard, and T. F. Spence. 1971. The functional and classificatory significance of combined metrical features of the primate shoulder girdle. J. Zool. Lond. 163:319-350.

Ashton, E. H., R. M. Flinn, C. E. Oxnard, and T. F. Spence. 1976. The adaptive and classificatory significance of certain quantitative features of the forelimb in primates. J. Zool. Lond. 179: 515-556.

Ashton, E. H., M.J.R. Healy, and S. Lipton. 1957. The descriptive use of discriminant functions in physical anthropology. Proc. Roy. Soc. Lond. Ser. B, 146:555-572.

Ashton, E. H. and C. E. Oxnard. 1958. Some variations of the maxillary nerve in primates. Proc. Zool. Soc. Lond. 131:457-470.

Ashton, E. H. and S. Zuckerman. 1950. Some quantitative dental characters of fossil hominoids. Phil. Trans. Roy. Soc. Lond. 234B:485-520.

Ashton, E. H. and S. Zuckerman. 1954. The anatomy of the articular fossa (fossa mandibularis) in man and apes. Amer. J. Phys., Anthrop. 12:29-61.

Ashton, E. H. and S. Zuckerman. 1958. The infraorbital foramen in the hominoidea. Proc. Zool. Soc. Lond. 131:471-485.

Bronowski, J. and W. M. Long. 1952. Statistics of discrimination in anthropology. Amer. J. Phys. Anthrop. 10:385-394.

Broom, R. and G.W.H. Schepers. 1946. The South African fossil ape-man. The australopithecinae. Trans. Mus. Mem. 2:1-272.

Butzer, K. W. 1976. "Paleoecological and contextual for the australopithecines." Communication to Symposium: "Australopithecines and their relation to hominid evolution." Northern Illinois University, Dekalb, Illinois, April, 1976.

Cave, A.J.E. 1973. The primate nasal fossa. J. Linn. Soc. 5:377-387.

Clark, W. E. Le Gros. 1959. The Antecedents of Man. University Press, Edinburgh.

Corrucini, R. S. and R. L. Ciochon. 1976. Morphometric affinities of the human shoulder. Amer. J. Phys. Anthrop. 45:19-37.

Day, M. H. and B. A. Wood. 1968. Functional affinities of the Olduvai hominid 8 talus. Man 3:440-455.

Day, M. H. and J. Wickens. 1980. Hominid footprints, photogrammetry and bipedalism. Proc. Anat. Soc. Grt. Brit. Ire. 26:19-22.

Erikson, G. E. 1963. Brachiation in New World monkeys and in anthropoid apes. Symp. Zool. Soc. Lond. 10:135-163.

Feldesman, M. R. 1976. The primate forelimb: a morphometric study of locomotor diversity. Univ. Oregon Anthrop. Papers 10:1-154.

Fleagle, J. 1976. Locomotor behavior and skeletal anatomy of sympatric Malaysian leaf monkeys (Presbytis obscura and Presbytis melalophos). Yrbk. Phys. Anthrop. 20:440-453.

Friedenthal, H. 1900. Ueber einen experimentellen Nachweiss von Blutsverwantschaft. Arch. Anat. Physiol. Lpz., Physiol. Abt. 494-508.

Goodman, M. 1976. Towards a genealogical description of the primates. In: Molecular Anthropology. Eds. M. Goodman and R. E. Tashian. Plenum, New York, pp. 321-353.

Gould, S. J. and Eldredge. 1977. Punctuated equilibria: the tempo and mode of evolution reconsidered. Paleobiology 3:115-151.

Groves, C. P. 1974. Taxonomy and phylogeny of prosimians. In: Prosimian Biology. Eds. R. D. Martin, C. A. Doyle and A. C. Walker. Duckworth, London, pp. 449-473.

Hansinger, M. J. 1976. Hominids in plio-pleistocene Africa: an odontometric approach to phylogeny. Ph.D. diss., University of Florida.

Johanson, D. C. and M. Taieb. 1979. Plio-pleistocene hominid discoveries in Hadar, Ethiopia. Nature 260:293-297.

Johanson, D. C. and T. D. White. 1979. A systematic assessment of early African hominids. Science 203:321-330.

Jouffroy, F. K. 1975. Osteology and myology of the lemuriform postcranial skeleton. In: Lemur Biology. Eds. I. Tattersall and R. S. Sussman. Plenum, New York, pp. 149-192.

[Leakey, R.E.F.] 1971. Australopithecus, a long armed, short legged knuckle walker. Science News 100:357.

Leakey, R.E.F. 1973. Further evidence of lower pleistocene hominids from East Rudolf, North Kenya, 1972. Nature 242: 170-173.

Lewis, O. J. 1981. Functional morphology of the joints of the evolving foot. Symp. Zool. Soc. Lond. 46: 169-188.

Lisowski, F. P., G. H. Albrecht and C. E. Oxnard. 1974. The form of the talus in some higher primates. Amer. J. Phys. Anthrop. 41:191-215.

Lisowski, F. P., G. H. Albrecht and C. E. Oxnard. 1976. African fossil tali: further multivariate morphometric studies. Amer. J. Phys. Anthrop. 45:5-18.

Luckett, W. P. 1975. Ontogeny of foetal membranes and placentae: their bearing on primate phylogeny. In: The Phylogeny of the Primates. Eds. W. P. Luckett and F. Szalay. Plenum, New York, pp. 157-182.

Manaster, B.J.M. 1975. Locomotor adaptations within the Cercopithecus, Cercocebus, and Presbytis, genera: a multivariate approach. Ph.D. Thesis, The University of Chicago, Chicago.

Manaster, B.J.M. 1979. Locomotor adaptations within the Cercopithecus genus: a multivariate approach. Amer. J. Phys. Anthrop. 50:169-182.

McArdle, J. E. 1978. Functional anatomy of the hip and thigh of the Lorisiformes. Ph.D. Thesis, The University of Chicago, Chicago.

McArdle, J. E. 1981. Functional morphology of the hip and thigh of the lorisiformes. Karger, Basel.

McHenry, H. M. 1973. The post cranial skeleton of early Pleistocene hominids. Ph.D. diss., Harvard University, Cambridge, Mass.

McHenry, H. M. and R. S. Corruccini. 1975. Multivariate analysis of early hominoid pelvic bones. Amer. J. Phys. Anthrop. 46:263-270.

Minkoff, E. C. 1974. The direction of lower primate evolution: an old hypothesis revived. Amer. Nat. 108:519-532.

Mittermeier, R. A. 1978. Locomotion and posture in Ateles geoffroyi and Ateles paniscus. Folia Primatol. 30:161-193.

Mollison, T. (1910). Die Korperproportionen der Primaten. Morph. Jhb. 42:79-304.

Nuttall, G.H.F. 1904. Blood Immunity and Blood Relationship. Cambridge University Press, Cambridge.

Oxnard, C. E. 1967. The functional morphology of the primate shoulder as revealed by comparative anatomical, osteometric and discriminant function techniques. Amer. J. Phys. Anthrop. 26:219-240.

Oxnard, C. E. 1973a. Form and Pattern in Human Evolution: Some Mathematical, Physical and Engineering Approaches. University of Chicago Press, Chicago.

Oxnard, C. E. 1973b. Some locomotor adaptations among lower primates. Symp. Zool. Soc. Lond. 33:255-299.

Oxnard, C. E. 1975a. Uniqueness and Diversity in Human Evolution: Morphometric Studies of Australopithecines. University of Chicago Press, Chicago.

Oxnard, C. E. 1975b. Primate locomotor classifications for evaluating fossils: their inutiligy, and an alternative. In: Proc. Symp. 5th Congr. Internat. Soc. Ed. S. Kondo. Japan Science Press, Tokyo, pp. 269-284.

Oxnard, C. E. 1975c. The place of the australopithecines in human evolution: grounds for doubt? Nature 258:389-395.

Oxnard, C. E. 1977a. Morphometric affinities of the human shoulder. Amer. J. Phys. Anthrop. 46:367-374.

Oxnard, C. E. 1977b. Human fossils: the new revolution. In: The Great Ideas Today. Eds. R. M. Hutchins and M. J. Adler, Britannica Press, Chicago, pp. 92-153.

Oxnard, C. E. 1978a. Primate quadrupedalism: some subtle structural correlates. Yrbk. Phys. Anthrop. 20:538-554.

Oxnard, C. E. 1978b. The problem of convergence and the place of Tarsius in primate phylogeny. In: Recent Advances in Primatology, Volume 3, Evolution. Eds. D. J. Chivers and K. A. Joysey. Academic Press, London, pp. 239-247.

Oxnard, C. E. 1978c. One biologist's view of morphometrics. Ann. Rev. Ecol. Syst. 9:219-241.

Oxnard, C. E. 1979a. Some methodological factors in studying the morphological behavioral interface. In: Environment, Behavior and Morphology: Dynamic Interactions in Primates. Eds. M. E. Morbeck. H. Preuschoft and N. Gomberg. Fischer, New York, pp. 183-207.

Oxnard, C. E. 1979b. The morphological behavioral interface in extant primates: some implications for systematics and evolution. In: Environment, Behavior and Morphology: Dynamic Interactions in Primates. Eds. M. E. Morbeck. H. Preuschoft and N. Gomberg. Fischer, New York, pp. 209-227.

Oxnard, C. E. 1979c. The relationships of Australopithecus and Homo: another view. J. Hum. Evol. 8:427-432.

Oxnard, C. E. 1980. Convention and controversy in human evolution. Homo 30:225-246.

Oxnard, C. E. 1981a. The place of man among the primates: anatomical, molecular and morphometric evidence. Homo 3:23-53.

Oxnard, C. E. 1981b. The uniqueness of Daubentonia. Amer. J. Phys. Anthrop. 54:1-22.

Oxnard, C. E. 1982. The Order of Man: a Biomathematical Anatomy of the Primates. Hong Kong University Press, Hong Kong.

Oxnard, C. E., R. German, F.-K. Jouffroy and J. Lessertisseur. 1981. The morphometrics of limb proportions in leaping prosimians. Amer. J. Phys. Anthrop. 54:421-430.

Oxnard, C. E., R. German and J. E. McArdle. 1981. The functional morphometrics of the hip and thigh in leaping prosimians. Amer. J. Phys. Anthrop. 54: 481-498.

Oxnard, C. E. and F. P. Lisowski. 1980. Functional articulation of some hominoid foot bones: implications for the Olduvai (Hominid 8) foot. Amer. J. Phys. Anthrop. 52:107-118.

Patterson, B. and W. W. Howells. 1967. Hominid humeral fragment from early pleistocene of northwestern Kenya. Science 156:64-66.

Priemel, G. 1937. Die platyrrhinen Affen als Bewegungstype unter besondere Berucksichtigung der Extremformen *Callicebus* und *Ateles*. Z. Morph. Okol. Tiere. 33:1-52.

Rak, Y. and R. J. Clark. 1979. Ear ossicle of *Australopithecus robustus*. Nature 279:62-63.

Robinson, J. T. 1972. Early Hominid Posture and Locomotion. University of Chicago Press, Chicago.

Rodman, P. S. 1979. Skeletal differentiation of *Macaca fascicularis* and *Macaca nemestrina* in relation to arboreal and terrestrial quadrupedalism. Amer. J. Phys. Anthrop. 51:51-62.

Schultz, A. H. 1969. The Life of Primates. Universe Books, New York.

Senut, B. 1981. Humeral outlines in some hominoid primates and in pliopleistocene hominids. Amer. J. Phys. Anthrop. 56:275-284.

Simpson, G. G. 1953. The Major Features of Evolution. Columbia University Press, New York.

Szalay, F. 1975a. Phylogeny, adaptations and dispersal of the tarsiiforme primates. In "Phylogeny of the Primates." Eds. W. P. Luckett and F. Szalay. Plenum, New York.

Szalay, F. S. 1975b. Phylogeny of primate higher taxa. The basicranial evidence. In: The Phylogeny of the Primates. Eds. W. P. Luckett and F. Szalay Plenum. New York, pp. 91-125.

Tobias, P. V. 1967. The cranium and maxillary dentition of *Australopithecus* (*Zinjanthropus*) *boisei*. In "Olduvai Gorge, Volume 2." Ed. L.S.B. Leakey. Oxford University Press, Oxford.

Wood, B. A. 1973. A *Homo* talus from East Rudolf, Kenya. *J. Anat. Lond*. 117:203-24.

Woo, Ru Kang. 1981. Recent advances in Chinese palaeoanthropology. *Hong Kong University Press Occasional Papers* 2:1-32.

Zuckerman, S. 1933. *Functional Affinities of Man, Monkeys and Apes*. Kegan Paul, London.

Zuckerman, S., E. H. Ashton, R. M. Flinn, C. E. Oxnard, and T. F. Spence. 1973. Some locomotor features of the pelvic girdle in primates. *Symp. Zool. Soc. Lond*. 33:71-165.

A BEHAVIORAL RECONSTRUCTION OF AUSTRALOPITHECUS

Adrienne L. Zihlman
University of California, Santa Cruz

Becoming Hominid: The Critical Questions

Reconstructions of early human behavior are necessarily speculative. Nevertheless, social life is shaped by the biology of the actors and by how and what foods are obtained, so that it is possible to formulate plausible theories. For most of human evolutionary history, until the last 10,000 years, humans moved from place to place, carrying their few possessions, collecting and sharing widely dispersed plant and animal foods--a gathering-hunting way of life that contrasted with sedentary life which emerged with food production. In traditional reconstructions of the early hominid life style, gathering is treated as incidental; hunting and meat-eating are depicted as dramatic innovations that define and mold human behavior, distinguish it from the behavior of the ape ancestors, and culminate in the emergence of *Homo*.

During the last few years, much has been written on the life ways of early hominids. Some authors pursue the issue of hunting and take little notice of gathering (Wilson 1980, Campbell 1982, Mellen 1981). Others incorporate gathering with hunting and emphasize the "mixed economy" (Leakey 1981, Isaac 1978). And others continue to emphasize that the hominid diet, like that of chimpanzees and other primates, was omnivorous, and meat-eating not a distinguishing human characteristic (Harding and Teleki 1981, Zihlman and Tanner 1978). Some authors have set up a false dichotomy between "man the hunter" and "woman the gatherer" (e.g., Washburn 1981). The critical questions are often not addressed: what behaviors emerged between three and four million years ago? If gathering, from what behavioral precursors did it emerge? If hunting, what preceded it? On what evidence are our hypotheses to be formulated, and how are our conclusions to be tested? If gathering preceded hunting, what are the implications for interpretations of human evolution?

I review recent evidence to test the "gathering model" (Tanner & Zihlman 1976), a model which I believe still stands as a viable hypothesis for this *earliest period* of human existence. By emphasizing that *both hunting* and gathering are new and arose together (e.g.,

R.E.F. Leakey 1981), the question of how hunting developed remains unaddressed. In what follows, I argue that the way of life of early hominids was based neither on hunting, nor even gathering and hunting, but on an omnivorous diet obtained by collecting and carrying plant foods, and by catching small animals without tools. Because of the long dependency of the young after weaning, to master tool making and food-getting skills, food was shared by adults with juveniles. Social innovations such as sharing are as integral to the human way of life as tools or bipedal locomotion. This primarily "gathering" way of life--though it included some meat-preceded hunting with tools by _at least_ one-and-one-half million years.

In this brief overview, I deal with the time frame between two and four million years ago. I refer collectively to the early hominids living then as _Australopithecus_, including _A. afarensis_ and _A. africanus_, because they share many anatomical characteristics (Tobias 1980). They are differentiated as a group from the apes by bipedal locomotion, large grinding teeth and smaller canines, but the Australopithicines had chimpanzee-sized brains. In all probability a gracile australopithecine species, (perhaps _A. afarensis_) gave rise to _Homo_. By two million years ago, there are two genera of hominids, _Homo_ and a robust _Australopithecus_, stone tools and evidence of animal butchering. Although this later evidence plays little role in reconstructing the prior two million years, its implications for interpreting the early period will be discussed briefly.

Evidence and Its Interpretation: Continuity in Evolution

In a court of law, some evidence is thought to be so clear that "it speaks for itself" (_res ipsa loquitur_), but in anthropology, an analytical framework is needed for the interpretation of data. That framework is evolution. Individuals differ in their ability to survive and reproduce in a particular environment, and this ability changes as the environment changes. Therefore, it is essential to understand the environmental context for human evolution. Interpretations of human evolution also must include discussion of females, as well as males, and infants and young, as well as adults. Infant survival to adulthood is critical to species survival, and females are central not only to producing offspring, but to socializing and caring for them, thus ensuring their survival.

Evolution is and has been a continuous process, whether or not the record of the past reflects this precisely. It is in the recombination of existing behavioral elements that a new way of life is initiated; this demonstrates that evolution is also behavioral change. Often the fossil evidence appears discontinuous, but there must have been behavioral continuity during all phases of human evolution. Furthermore, continuity in evolution assumes as integrated way of life, and to appreciate that fully, it is necessary to explore interrelated patterns, which include social, technological, economic and ultimately ideological features.

The record of the past includes the fossils themselves, associated findings of other species, evidence of hominid activity, and the ancient ecological setting. Because the fossil record does not stand alone (Kitts 1974), to interpret the record and reconstruct a way of life requires supporting evidence from observations on closely related living groups--namely, nonhuman primates and gathering-hunting peoples (Lee 1979, Marshall 1976, Silberbauer 1987, Gale 1974, Lee & DeVore 1976, Dahlberg 1981, Bicchieri 1972). Using the past and present evidence, we may deduce something about body size, locomotor systems, habitat and possible diet, potential predators and the circumstances of death.

From anatomy, behavior, and protein structure and DNA, it has been deduced that man, chimpanzee and gorilla diverged from a common ancestor as recently as five or six million years ago (King and Wilson 1975, Sarich and Cronin 1976). No hominid fossils older than this have been found. The oldest fragmentary remains of a jaw and an arm bone from Lothagam and Kanapoi, respectively, in northern Kenya, are dated about five million years (Coppens et al., 1976). The most complete evidence includes footprints, pelvic and limb bones, and comes from sites in Ethiopia, the Hadar Formation, in Tanzania at Laetoli, and in South Africa at Sterkfontein cave; these sites date between 2.5 to over 3.5 million years ago (Howell 1978).

The fossil sites of eastern and southern Africa occur in what was savanna mosaic, not forest. The abundant fossils from this time period are mostly jaws and teeth, though a dozen or more skulls have been found, and almost all parts of the skeleton. The earliest hominids were small, about the size of chimpanzees, but having a different body build. Their teeth, too, are distinctive: small canines and large cheek teeth with thick enamel, usually showing extensive wear. Cranial capacity ranged from somewhat less than 400 cm^3 to less than 500 cm^3. See Jolly (1978) for detailed discussion.

Pygmy Chimpanzees as a Model for a Hominid Precursor

The close genetic affinity between humans and African apes and their recent time of divergence--a three-way split about five million years ago--links the origin of hominids to that of chimpanzees and gorillas (Sarich & Cronin 1976). Therefore, a morphological model for the proposed common ancestor must be suitable for all three species. Of the African apes, the pygmy chimpanzee (*Pan paniscus*) has been proposed to most closely represent, anatomically and behaviorally, the ancestral species, just prior to the divergence and radiation of the hominids (Zihlman et al., 1978).

Pygmy chimpanzees, named on the basis of their smaller skulls and dentition compared to the common chimpanzee (*Pan troglodytes*) are by no means "pygmy" on the basis of their limb bones. They are a distinct species, and more generalized than the common chimpanzee (Coolidge 1933, Zihlman and Cramer 1978) in these respects: lighter body weight, shorter upper than lower limbs, little overall sexual dimorphism, especially in cranial capacity and canine teeth. Compared to early hominids, pygmy chimpanzees are similar in overall size as indicated by stature (femoral length), joint sizes, and cranial capacity. Postcranial differences relate mainly to bipedal locomotion: shorter upper limb and innominate length in *Australopithecus*, as well as greater iliac and sacral breadths (Zihlman 1979). Other traits in which pygmy chimpanzees and *Australopithecus* differ--larger and more thickly enameled molars and possibly a greater degree of sexual dimorphism--can be viewed as adaptations to savanna living.

Information on social behavior of pygmy chimpanzees is becoming available and offers the opportunity for refining this model and testing previous hypotheses about the relation of sexual dimorphism to mating patterns, frequency and function of food sharing, and nature of social interaction and group structure. For example, pygmy chimpanzees seem the most social of the apes: most affiliative behavior is between the sexes, next, between females, and least frequent between males--

a pattern which contrasts with that in the common species in which interaction is most frequent among males and least among females (Kuroda 1979). Group structure is larger, on average, than in the common chimpanzee species; solitary individuals are rare, and groups, regardless of size, usually contain at least one male and one female--in contrast to the common species where all-male groups and solitary animals are observed more frequently (Kuroda 1980, Kano 1980). Of all the apes, pygmy chimpanzees have the smallest canines relative to body size (Leutenegger & Kelly 1977), and this anatomical feature may correlate with their greater degree of sociality.

Information of pygmy and common chimpanzees helps us to see the continuity between ape and human, to imagine how an ape can evolve into a hominid, and to remind us that the gap is narrow indeed. Comparison of pygmy chimpanzees and early hominids enables one to reconstruct the sequence of morphological changes that occurred in the transition from a quadrupedal, knuckle-walking, climbing ape, to a structurally and behaviorally adapted bipedal hominid. If the ancestral form were similar to _Pan paniscus_ in body build, then during the shift from quadrupedal to bipedal locomotion, one of the earliest changes in body build might have been reduction in upper limb length and weight relative to the rest of the body, serving to lower the center of gravity and making upright posture more stable. The degree of morphological difference between pygmy chimpanzees and early hominids is less than that between _Australopithecus_ and _Homo sapiens_. We know the latter transition spanned a period of 3.5 million years. Therefore, it is conceivable that the transition from a chimp-like ancestor to early hominids occurred in less than three million years.

Delineating Assumptions

I want to be clear about my approach to reconstruction and the assumptions on which it is based. First, I assume that hominid origins must be consistent with relationships between social organization and ecology established for other species. I assume, too, that the new savanna environment was distinct from the forest habitat of the ape ancestors and required a new combination of behaviors ranging from locomotor to social. Because of the genetic closeness of the African apes to humans, I look to them for "elements" or "precursors" in behavior that can be seen to be important in early hominids.

Finally, because of the importance of demonstrating continuity between ape and human, I assume that if a behavior occurs in chimpanzees or gorillas (and even in other primates in some instances) as well as in contemporary gatherer-hunters, then the behavior must have been present among early hominids. This assumption, I maintain, establishes reconstructions on firm terrain and provides a number of "givens." For example, early hominids are characterized by an omnivorous diet (Harding and Teleki 1981); female mobility is a fundamental primate characteristic (Zihlman 1981); early hominids are social and there are long-term ties between mother and infant and among siblings.

A specific example of erroneous assumptions concerns the much debated issue of tool-using and making, particularly those manufactured from stone. Given the fact that chimpanzees use a wide variety of tools, primarily of organic materials (Goodall 1968), and that women and men who gather food use implements made from organic materials (Lee 1979), it is probable that early hominids did make extensive use of organic tools for digging and carrying long before the "invention" of stone implements as they are known in the archeological record. If early hominid females three million years ago used baskets of bark to collect plant foods, the record would leave no trace. It is unlikely that stone tools were invented without a prior tradition of perhaps several million years of tool-using (Lancaster 1968). Nonetheless, Lovejoy (1981) maintains that hominid origins and bipedalism came about long before hominids used tools--and only stone tools count as tools.

Stressing continuity in evolution helps us to avoid "bending the rules" for interpreting early human behavior. The early hominids were primates; they were nomadic and savanna-living, not a species so totally apart from all others that it is impossible to understand their way of life. At the same time, the australopithecines were not chimpanzees, nor were they exactly like gatherer-hunters today. But knowledge of the behavior of African apes, of savanna-living nomadic peoples, and of the savanna mosaic can reveal how the environment poses constraints on behavior, and how systems of behavior interrelate. The shift from ape to hominid involved a complex of interrelated behaviors, not just diet or sharing or bipedalism. The adaptation grew out of, but had its roots in, a forest-living chimp-like ancestor. Initial success on the African savanna preceded by some three to four million years the expansion of hominids to the world stage of temperate and frigid, as well as tropical environments.

Hominid Origins in Africa: Expansion into the Savanna

There are no fossils at all for the living African apes, but the earliest hominid fossils, *Australopithecus*, occur in a savanna environment in Africa about five million years ago (Howell 1978). Savannas consisted then, as now, of a mosaic of vegetation, not simply a broad expanse of grassland--our usual image--but of tall grasses, low bushes, and wooded areas around lakes and rivers. It is a patchy and fluctuating environment, and the distinct seasonality of wet and dry seasons is a major feature (Bourliere and Hadley 1970). The savannas of eastern Africa became widespread in the later part of Miocene, initiated by the formation of the Rift Valley from the collision of the African plate against Eurasia (Laporte and Zihlman, in preparation). The once continuous tropical forest, which extended through central Africa from the west to east coast, was thereby cut off in eastern Africa. The spreading habitat--the savanna mosaic--offered opportunities for species to expand into and adapt to it. Indeed, the fossil record shows a savanna radiation of pigs, bovids and monkeys, in addition to hominids (Maglio and Cooke 1978).

Gathering as the Earliest Innovation

The savanna mosaic offered abundant plant and animal food to an omnivorous ape; the exploitation of these resources can best be imagined as a way of life involving the gathering of food as the earliest innovation, as opposed to hunting with tools. The innovation was not so much a change in the kinds of foods eaten, but in the combination of bipedalism and using tools to obtain it. Many savanna foods are underground--roots and tubers--hidden away from the ravages of seasonal drought and heat; nuts and seeds are protected in thick outer shells; standing water is less available during some seasons; many plant foods are highly seasonal and widely dispersed; fruit is less available than in forested habitats.

Gathering means collecting food, primarily plants, but also small animals, generally dispersed over a wide area, occasionally concentrated in one place, thereby departing from the ape pattern of individual foraging and eating where food is found. Gathered food is obtained in quantities larger than those that one individual would eat on the spot and conveyed to a protected site where it can be prepared or cleaned and shared with other members of the group. Tools such as containers and digging sticks are a great advantage for gathering: containers permit the collection of large quantities of small items such as seeds, berries, nuts, roots and insects, so that potential food sources are not limited by their size or shape. Digging tools would give access to roots and tubers. Wooden or stone tools would aid preparation of plant foods through scraping, pounding, cracking, or cutting. A sling for the baby, a container, and digging tools would have been minimal requirements for a mother gatherer, ensuring her mobility and enabling her to obtain a wide range of foods quickly and bring them back to a safe place for preparation, sharing and consumption.

Behavioral preadaptations for gathering that most likely existed in the ancestral ape population as hypothesized from our knowledge of chimpanzees include: 1) the ability to carry objects, both food and tools, 2) the use of tools for obtaining unseen foods and for preparation, 3) sharing between mother and offspring and between males and females, and 4) occasional bipedalism. But what was possible and occasional in the ape ancestors became regular and intensive in the hominids and eventually evolved into the food-producing and highly technical way of life that is so distinctively human.

Gathered plant foods, insects and small vertebrates probably accounted for more than 90% of early hominid diet. However, compared with apes, meat consumption no doubt increased through predatory behavior. Because of the greater availability of young and small animals from the numerous herd species on the savannas (Bourliere 1963), the frequency and sophistication of predatory activity probably increased, similar to what happened in a troop of savanna-living baboons at Gilgil, Kenya, due to an increase in the antelope population (Harding & Strum 1976). For this early period between two and four million years ago, scavenging and consumption of large dead animals found by chance were probably infrequent activities. This meat source apparently did not become available until the invention of stone cutting tools and butchering. There is no evidence for stone tools associated with large animals for this earliest time period. But regular protein sources could have come from eggs, insects and other invertebrates, and vertebrates obtained by using gathering techniques.

Adaptations for Life in the Savanna

The gathering hypothesis must be evaluated in terms of australopithecine anatomy: pelvis and foot, dentition, hand, brain and body size. These features must be viewed not only as separate systems, but as interrelated systems. For example, locomotor behavior is intimately related to getting food, avoiding predators, and encountering potential mates. And, the time required for the young to develop the adult pattern has profound implications for parental care.

Bipedalism: Commitment to the Savanna Habitat

Bipedal locomotion is a terrestrial, savanna adaptation for covering long distances while carrying food and water, digging sticks, objects for defense, and offspring. It is a versatile form of locomotion, but is not an adaptation for speed; rather, endurance is its hallmark. Often, bipedalism has been viewed as a necessity for tracking animals, for carrying hunting tools and meat back to camp. But getting food on the savanna must have required both sexes moving over longer distances than the ape ancestors. Endurance for long-distance walking does not differ by sex. The large home range needed for obtaining the widely dispersed resources on the savanna suggests that both females and males traveled frequently and far. With hands and arms freed from locomotor functions, effectiveness of tool-using, of capturing small prey with the hands, and of dealing with predators, must be enhanced also.

The evidence for hominid bipedalism appears earliest from Laetoli, Tanzania, where Mary Leakey discovered footprints in volcanic ash dated to some 3.5 million years ago (M. D. Leakey & Hay 1979). The footprints are those of a hominid, not a pongid, with the great toe more in line with the other toes than is found in apes. By three million years ago, further evidence from fossil pelves, knee joints and foot bones from Hadar, Ethiopia and Sterkfontein, South Africa, confirm bipedalism (Johanson & Taieb 1976, Robinson 1972, Le Gros Clark 1967). The discovery of "Lucy" (Al 288 from Hadar) shows that body build differed from both modern chimpanzees and humans (Johanson & Taieb 1976, Zihlman 1979, 1982).

Bipedal locomotion and its anatomical basis has long defined the hominid family--Hominidae. The structural basis for effective bipedal locomotion includes changes in the muscles and bones--in their shape and position and relative weight (Zihlman & Brunker 1979). Feet cannot perform two competing functions simultaneously and effectively: mobility and stability. A mobile foot, as in chimpanzees, is effective for climbing and moving through trees, but contrasts with the hominid foot that is designed for weight-bearing: less mobile ankle and foot joints, shorter toes, larger and heavier foot bones (Zihlman & Brunker 1979).

The pelvis at 3 million years resembles chimpanzees in the size of the joints (sacroiliac, hip and lumbosacral) but differs in overall shape in having a shorter, broader ilium and broader sacrum (Zihlman 1979). Size, shape and position of the sacrum indicate the presence of a lumbar curve, and consequently an upright trunk. The shape and size of the ilium suggest a change in size and position of the hip joint muscles from those of the ape ancestor, particularly those muscles rotating the body forward over the feet.

Social aspects of bipedalism are rarely emphasized in evolutionary reconstructions. Continuous with the chimp-like ancestral pattern, mobility of early hominid mothers must have been maintained, and the young hominids required several years for developing independent locomotion. Among chimpanzees, an infant takes its first steps at about six months of age, but steady locomotion does not occur until the third year; young chimpanzees ride on their mother's back until the fourth year (Goodall and Hamburg 1975). The foot, once changed for bipedalism, necessitated new ways of carrying babies. Without a grasping foot like chimpanzees, which enables babies to cling to their mother's hair, hominid babies could not cling to their mothers, who probably had less hair to hold on to, and the mother had to take a more active role in carrying. A hominid mother probably carried her baby almost continually for three years, and occasionally for another two or three. In modern humans, a baby's weight adds stress to its mother's hip, knee and ankle joints even before birth. Mother-infant ties would have been strong, with the infant even more dependent upon its mother than among chimpanzees. Because of the demands and consequences of bipedalism, Linton's suggestion (1971) that a sling to support an infant as an early hominid "tool" seems very reasonable. This tool would have freed the mother's arms and hands for obtaining food, using tools, and carrying food.

Dentition

Australopithecine teeth differ from both contemporary apes and humans. Relative to body size, the thickly enameled premolars and molars are larger than those of apes or humans, and are often worn and chipped even in young individuals (Wolpoff 1973; Wallace 1978). These features are associated with a large surface area for grinding foods that are tough and require a great deal of chewing, or foods from the ground that are gritty and wear down the teeth. The skull and face, shaped to a considerable extent by the muscles of mastication, show well-developed temporal and masseter muscles. The dental and cranial features, then, provide evidence for an omnivorous diet of gritty and tough foods that may have included fibrous vegetation, seeds, roots, and other food from the ground.

The australopithecine canine teeth are smaller than those of any living ape and show little sexual dimorphism (Wolpoff 1975). Among primates, large canine teeth--larger in males than females--are often associated with male/male competition and with defense against predators (Leutenegger and Kelly 1977). The small hominid canines might be interpreted as indicating little male/male competition, or an alternative means of defense against predators than large canines.

Body Size and Sexual Dimorphism

Most parts of the early hominid skeleton are comparable in size to those of chimpanzees with a body weight between 25 and 60 kg (Zihlman 1976; 1979). The extent to which variability in size of isolated parts of the skeleton can be attributed to sexual dimorphism or to possible species differences is not entirely clear. The variability discussed with the species Australopithecus afarensis has been attributed to an extreme degree of sexual dimorphism (Johanson and White 1979). Some species of primate show such marked sexual dimorphism; the male weighs much more than the female, and this weight difference is reflected in their body size dimensions. In moden humans, and in pygmy and common chimpanzees, body weight difference between the sexes is moderate--not extreme -- and dimensions such as long bone lengths and cranial capacity between the sexes of each species, may or may not vary by sex (Zihlman 1976, Cramer and Zihlman 1978).

What can we say about Australopithecus in the absence of complete information? Little variation can be detected in cranial capacity or in size of canine teeth, but perhaps at least moderate variation in body size existed (Zihlman 1976). Whether the variability among early hominids is due to species differences or to sex differences, the configuration of traits usually expressing sexual dimorphism is unlike any pattern found among living hominoids.

Hand Bones: An Early Tool User

During this earliest period of human evolution, there is no evidence for stone tools in the record whereas the evidence for bipedal locomotion is clear. Hands relate most obviously to tool using, but also to locomotor behavior. In a bipedal species, hands are no longer used for supporting body weight. As with feet, hands cannot be equally effective in weight-bearing and in manipulation, because the underlying structural bases differ.

All the great apes have big, stong hands, with long, curved, heavy phalanges and joints in the hand and wrist adapted for supporting their body weight in quadrupedal walking. Human hands have shorter, straighter fingers with greater movement possible between the joints in the palm. Mary Marzke who has studied ape and human hands, as well as the Hadar hand bones, concludes from her study that Australopithecus afarensis had grips capable of manipulating objects in cutting, scraping, hammering, throwing, clubbing and digging against resistence (Marzke 1982).

If the origin of the hominid line meant a reliance on regular tool-making and tool-using (Washburn 1960), -- and we can argue about what the tools were made of and what they were used for -- then changes in hand structure may have occurred as early as changes in the foot and pelvis.

Brain and Behavior

Estimates of australopithecine cranial capacity from Hadar, Sterkfontein and Makapan are less than 400 cm^3 to about 450 cm^3 (Holloway 1975; Kimbel and White 1980). These estimates lie with the chimpanzee range, though on the large end of the spectrum. Because body size seems to be within the range of chimpanzees, only a very small increase in brain size relative to body size could have occurred at this earliest stage of hominid evolution. By two million years ago, there were larger brains, outside the size range of the apes (Tobias 1975); this later increase perhaps reflects the integration and refinement of hand skills, tool-using and associated cognitive and motor behaviors necessary to conceive and execute them.

Increase is only part of the picture. Considerable reorganization of the brain could have occurred even at this early stage in human evolution, and the integration of changing functions is as critical as increased number of neurons. The complexity of problem-solving and communication is likely to have exceeded that exhibited by apes. These abilities would have great survival value in enhanced efficiency of locating and extracting food by knowing the topography of the land, the seasonality of resources, and the locations of raw materials for making tools. Cognitive abilities would have been important for generalizing into new contexts those established patterns of tool-using, and into more complex social interactions involving food-sharing and newly learned communication skills.

Growing Up Australopithecine: Life Processes

Success by evolutionary standards is reproductive success: the transmission of the most genes, and therefore the most offspring, to the next generation. This is measured, not only by the actual number of offspring conceived and born to an individual, but also by the number of offspring which reach reproductive age, and so pass on their genes. A unique aspect of hominid survival was based on sharing food with young, especially after weaning, and cooperation in care and protection of young. Within the security of the social group, the young developed locomotor independence, skills in using and making tools, behavior patterns requiring time, so that the length of dependency on adults was necessarily extended. Social ties were strong and lasting.

Birth and Early Survival

Because the brain was still quite small, and the pelvic opening adequate, the birth process was probably relatively unproblematic compared to *Homo sapiens* (Leutenegger 1972). Hominid babies did not have a grasping foot so that mothers had to take a more active role in carrying babies, compared to chimpanzees. Early survival depended heavily upon a mother's biological and social fitness, her maintaining mobility and effectiveness at foodgathering while carrying an infant (Zihlman 1978).

Development and Learning

The fossil record, showing that *Australopithecus* dental development was comparable to the human pattern (Mann 1975), supports the hypothesis that a longer period of time was required to learn and acquire biomechanical, motor and social skills for this way of life. From a child's point of view, the gathering way of life required a great deal of learning and a long dependence on adults, which may have lasted for at least eight years. An australopithecine child could not be physically independent until it could walk long distances, master the skills of collecting and processing food with tools, and acquire sufficient knowledge of the physical and social environments.

It takes a long time, perhaps six to eight years, to develop the strength, endurance, adult body build, and physiology requisite for locomotor independence. A contempory human child can keep its balance and walk very short distances by three years of age, but the stamina to walk long distances does not develop until almost eight years of age. Among the !Kung of Kalahari in southern Africa, children are not recruited into subsistence because they lack the endurance to keep up on the march, need water, and want to be carried (Draper & Cashdan 1974).

In the realm of tool-using, presumably it would have taken an australopithecine child at least the five years to learn motor skills that it does for young chimpanzees to master the art of termiting with tools. If food was located underground and required preparation, then the skills to recognize environmental clues on where to dig, the strength to dig, and the making and using of tools for food preparation, all had to be learned. Because the young were carried from place to place on their mother's bodies they had a chance to watch her. So, within the protection of mother and the social group, young australopithecines would have had an extended time for learning, through observation and practice, the conceptual skills for understanding their environment, practical skills for survival, and social skills for communicating with other group members.

Weaning and Birth Spacing

In comparison with foraging for fruits in the forest, as chimpanzees do, gathering food on the savanna placed another burden on hominid females with young: each offspring had to be fed and tended for several years after weaning, until it acquired motor skills and knowledge. An extended nursing time to at least three years, as among modern gatherer-hunters (Howell 1979), presumably was followed by a long time of feeding by the mother and perhaps by other social group members. Mothers may have premasticated food for young before weaning, and carrying water for all young might have been a critical factor in survival. Savanna foods are generally rough and would have been difficult to digest, especially in the absence of fire; for the early time period, there is no evidence for use of fire.

In contrast with modern gatherer-hunters, there is no evidence for early campsites. This could mean that older children were not left behind while mothers gathered, or perhaps there were no adults available to care for them, as all were involved in getting food.

There was extensive maternal investment: in carrying, nursing, processing food to feed after weaning, birth spacing to accommodate the mother's workload of carrying babies and food and water, and in giving each child psychological security during development. Given the importance of a mother's mobility, it is unlikely that she would increase her workload by carrying two infants and still gather food (see Lee 1979). Therefore, a child's survival would also depend upon its mother's ability to feed it after weaning, and to space her infants, so that each might receive the care necessary for its survival to adulthood.

Mating and Sexual Behavior

In the "hunting scenario" of male/female relations, the human female lost estrus so as to be perpetually receptive to her male protector and provider, so pair bonds were formed. In a more recent version of this type of mating system, Lovejoy (1981) postulates that males were the sole food providers -- though not of meat from the hunt (new twist); they formed pair bonds in order to ensure each had an available sexual partner.

Mating patterns were likely "polygamous" rather than "monogamous" for several reasons. First, there appears to be a positive correlation between maternal investment and female choice of mates (Trivers 1972), and no mothers invest more time and energy in their offspring than humans. And, there is evidence that female choice exists in the sexual behavior of apes and monkeys (Tutin 1979, Taub 1980). Second, in mammals generally and nonhuman primates in particular, body size differences between males and females correlates with mating systems in which some males, but no females do not reproduce -- a "polygynous" system (Leutenegger 1977, Zihlman 1976). Finally, hominids are a savanna species, where mobility is a fundamental part of the adaptation. Pair bonds are more common in mammalian (and primate) forest living species with a small home range than in savanna species with a larger one (Bourliere 1973).

Only conjecture is possible in reconstructing sexual activity. But it is difficult to imagine early hominids as having exclusive sexual partners, as Lovejoy has proposed, for anything but short periods of time. Exclusivity of sexual partners is rare in chimpanzees or gorillas, though there are individual preferences, as there are in other species of primates. And, even though the label of the Kalahari gatherer-hunters might be "monogamous," they may marry several times during their life, and even if married, often have lovers (Shostak 1981).

The evolving human pattern, with long dependency even after weaning, would have put a premium on male willingness to share and care for the young. In the early stages of hominid evolution this behavior, not tied to biological fatherhood, may have taken the form of helping with offspring of female friends and relatives--mother, aunts and sisters -- in the context of the social group. Males with these traits may have been viewed as sexually attractive by females (from other social groups) in choosing their mates and thereby sexual selection operates. The small canine teeth in *Australopithecus* suggests reduced aggressive behavior among males, and between female and male, in comparison with chimpanzees and gorillas.

Predation and Death

The fossil record for this early period gives little evidence on average life span. There is some suggestion, however, that australopithecines did not live until a "ripe old age." On average, most australopithecine individuals died in their early twenties, not long after acquiring their third molars (Mann 1975). Although not preserved well in the fossil record, infant mortality was probably high. Many hominid fossils show evidence of carnivore chewing which could indicate 1) the hominids died a natural death and were scavenged, or, 2) were killed and eaten by carnivores. The young, after weaning, may have been particularly vulnerable. There is evidence from Swartkrans cave that a nine-year-old australopithecine received a lethal puncture wound from the canine teeth of a leopard (Brain 1970, 1981).

Evidence of burials and ritual comes much later in time. Yet even chimpanzees recognize death and react emotionally to it (Teleki 1973a), and a young chimpanzee may not survive when its mother dies, even though it had been weaned (Goodall 1967). The increasing emotional interdependence among the early hominids would have increased the trauma of sudden death of a mother or sibling. The increasing conceptual abilities and self-awareness of early hominids were becoming integral to the hominid way of life.

Features of Australopithecine Communities

I approach the reconstruction of early hominid communities, from a comparative perspective, by noting the behavior of living African apes and moden humans which have been separated from each other about five or six million years, and each separated from the australopithecines about three million years. The patterns of dietary behavior, tool-using and social behavior do show continuity between ape and human (Teleki 1974, 1975). My reconstruction, therefore, is based upon what seems consistent with the ecological context of the savanna and what we know of chimpanzee and human nomadic groups.

Home Range

The African savannas, where *Australopithecus* lived and is found as fossils, support a diversity of plant species, animal species feeding on them and carnivores feeding on them. The hominids, as omnivores, should not be viewed as an open grasslands species, but as "ecotonal," utilizing resources throughout the savanna mosaic from the more forested areas along water courses, as well as grasslands and nearby hills (Behrensmeyer 1976, Butzer 1977). Because potential food sources were abundant, but scattered geographically and seasonally, the hominids would have had to move about a great deal; their home range must have been larger than any of the living nonhuman primates, including the savanna baboon and the woodland savanna chimpanzee. The hominid locomotor system, adapted for walking several miles a day for food and water and for carrying, enabled the group to cover several hundred square miles over several annual cycles.

A large home range and low population density were probably characteristic of australopithecine communities. These features are often attributed to a hunting way of life -- that a large range is necessary for supplying a hominid group with meat. A large range is as necessary for gathering plant foods, insects, eggs, as it is for hunting, although within this large range, small animals could be caught and eaten. Opportunistic plant feeders, such as eland, also have a large range and a low population density (Moss 1975). Even with little meat-eating and no hunting, early hominids were probably wide-ranging over the savanna.

Group Size and Organization

Perhaps the overriding feature of australopithecine groups might have been its "openness," characteristic to some degree of all great ape groups (Reynolds 1966). The flexibility in temporary and changing associations would have been necessarily in order to take advantage of environmental opportunities in obtaining food. Ecological constraints are only part of the picture. Early hominids, like chimpanzees and gatherer-hunters, were social and liked being together. Silberbauer (1981) discusses the disadvantages of the temporary associations of Kalahari gatherer-hunters; they are willing to tolerate the inconvenience in order to be together.

Some of the factors accounting for temporary associations of australopithecine groups might have been a limited number of suitable areas for sleeping, availability of abundant food, protection from predators, especially at night, and social needs -- for sexual contacts, renewal of friendships, opportunities for young to play together, sharing food and exchanging of information about the environment. A community, in the sense that we recognize it today, was probably beginning to develop even at this early stage in evolution.

I envision a group organization most resembling pygmy chimpanzees, where temporary associations are in groups of 10-40, and 67% of group compositions are mixed (Kano 1980). Perhaps early hominids had more "permanent" groups of about 10 individuals, consisting mostly of related individuals, but with some individuals, either male or female, that joined the core. This core might span 3 generations, and would join up with other such "permanent" groups at water sources or abundant food sources, or occasionally travel together. These smaller, but perhaps relatively stable groups, consisted of several ages and both sexes, with mother/infant ties strong, though not the only strong social tie.

The importance of the social group in defense should be stressed. On the open savanna, hominids were in more danger of being eaten by carnivores than their forest-dwelling ancestors had been. Lacking large canine teeth and grasping feet of apes, hominids must have relied on forms of defense other than fighting and retreating to trees. Their upright posture would have made them more visible to predators, but would also have given them better visibility for spotting danger. A group with several adult hominids could have formed an effective unit of defense if they used intimidation displays, like those of chimpanzees, and developed skill in using sticks and throwing stones.

Food Sharing and a Division of Labor

Food-sharing has long been noted as a feature of human, rather than nonhuman behavior (Washburn and DeVore 1961). It has become synonymous with hunting and a sexual division of labor (e.g., Isaac 1978). I believe that it is very likely that food-sharing emerged earlier and independent of hunting and a sexual division of labor, and that it grew out of several features existing in living chimpanzees and presumably found in the common ancestral population. Food-sharing does occur among nonhuman primates, particularly chimpanzees, the species best known for this behavior.

Food-sharing exhibits several characteristics among chimpanzees: both plant and animal foods are shared, and plants more frequently than animals; shared food is not a major source of nutrition, but seems to express social ties among adults and between mother and offspring (Silk 1978; McGrew 1975; Teleki 1973b). In sharing among pygmy chimpanzees, females are most frequently the donors giving food to juveniles and infants and other females (Kano 1980). The plant foods shared among pygmy chimpanzees include several classes: 1) those shared among adults were large enough to share and easy to eat, though difficult to find and highly valued; 2) those difficult to process were predominantly shared with immature individuals, though also with adults, and 3) items neither difficult to obtain nor difficult to process, and recipients were mostly juveniles and infants (Kano 1980).

Using this information to reconstruct a scenario, the origin of food sharing could be attributed to an outgrowth of several factors: 1) investment of mother in offspring; 2) social ties and friendly behavior expressed through food sharing, especially among adult females; 3) regularly eating savanna foods which require preparation, and are too difficult for young individuals to process themselves. Silk (1978) and McGrew (1975, 1979) suggest that foodsharing, especially of plants, between mother and infant may facilitate the transition from dependent suckling to independent foraging by infants. Among early hominids, a long period of development lengthened the time after weaning, before the young were able to obtain food on their own or to contribute it to the group. Thus, the transition after weaning would have been more critical in early hominids, and required more maternal care. Food sharing very early in human evolution may have expanded beyond the mother-infant bond to become part of the social matrix among individuals and between groups.

The development of regular sharing among early hominids might have been influenced by the kinds of food available on the savanna.If, for this early period, plant foods formed the bulk of the diet, as they do for modern gatherer-hunters (and for chimpanzees), then many plant species were probably underground or had hard coverings. Both of these classes of food would require cutting up for sharing, cleaning or removal of the hard covering. Chimpanzees have all the behavioral elements for dealing with such foods: extraction tools for obtaining ants and termites, clubs and stones for opening different species of nuts; occasionally carrying the nuts to a place for using a tool to open them, either from trees to ground or from one place to the next on the ground (Goodall 1968; McGrew 1979; Boesch and Boesch 1981).

Furthermore, adult females at Gombe engage more frequently than males in extracting termites (McGrew 1979). Adult female chimpanzees on the Ivory Coast more frequently than other age-sex classes utilize difficult techniques in opening some species of nuts (Boesch & Boesch 1981). If such behaviors -- including the propensity of females to share -- were present in the ancestral population, it seems a small step to imagine hominid mothers regularly obtaining savanna plant foods, preparing and cleaning them and sharing with offspring.

What does sharing suggest about a division of labor by sex? At this early stage of hominid evolution, it seems likely that such a division was not necessary. Both sexes were carried by their mothers and would learn about food and the environment from her, as reported for the Kalahari gatherer-hunters (Marshall 1976). Both sexes were probably doing much the same things, but females were more responsible than males for carrying and feeding the young. The dependence of the young on their mothers would seem to be more of a decisive factor in activity by sex, than sex _per se_. In gatherer-hunters today, mothers contribute 90% of child care, though men do love and interact with their children (Lee 1979), men contribute food and general protection.

Evidence for Meat-eating: the Role of Sharing and Hunting?

The earliest known evidence from the fossil/archeological record for meat-eating comes from Koobi Fora and Olduvai Gorge in East Africa (see review by Isaac and Crader 1981), dated to less than 2 million years ago -- a time range outside the scope of this paper. Because this evidence has been used to support the idea that hunting formed the basis for the way of life of early hominids, the evidence will be briefly discussed here.

Although no direct evidence, it is likely that meat was a dietary item of early hominids prior to 2 million years ago. As suggested earlier in this paper, both chimpanzees and baboons catch small animals and eat meat. And, given the higher density of hooved mammals on the savanna than in the forest, early hominids could have caught small or young ones fairly easily. This behavior need not have involved tools, or cooperation, and can be viewed as part of their opportunistic foraging pattern.

Recent research, using variety of techniques has established the presence of cutmarks on bones made by stone tools (Potts and Shipman 1981) thus providing evidence that butchering with stone tools occurred at Koobi Fora and Olduvai Gorge (Bunn 1981). Also micro-wear studies of early stone implements suggest they were used for cutting animal tissue and soft plant material, and for scraping and sawing wood (Keeley and Toth 1981). This research is indeed very exciting. However, these studies do not prove hunting--only meat-eating. Furthermore, this evidence does not refute the hypothesis that early hominid diet, technology, and sharing--prior to 2 million years--revolved primarily around gathering and processing plant foods.

The argument that meat-eating is tied to sharing and division of labor by sex (Isaac 1978) depends upon the presence of home bases. Binford (1981) has challenged interpretations of bone and stone accumulations from Olduvai Gorge as sufficient to conclude that these areas were used as home bases by hominids. Obviously social behavior is an integral part of hominid behavior and their adaptation to the savanna--whether or not it can ever be demonstrated in the archeological record. Sharing was critical but in its initial stages, I would argue, the focus was not upon sharing between adult males and females, rather the sharing of all foods between mothers and young--indeed the survival of the young depended upon it. Meat-eating, home bases and sexual division of labor do not necessarily precede sharing. The belief in the central role of hunting in human evolution dies hard, even though evidence of it diminishes, while that for gathering as the technical and social innovation upon which hunting is based increases.

Summary: the Australopithecine Adaptation

Descended from a chimpanzee-like, forest-living ancestor, _Australopithecus_ adapted to the savanna mosaic. Although, like their ancestors, early hominids were omnivorous, their techniques for obtaining and preparing food were new and involved increasingly sophisticated tools and social organization. Initially, australopithecines gathered and shared mainly plants, obtained with the aid of organic tools such as digging sticks and bark containers, and prepared with sticks or stones. Plant foods provided the staple, though meat was added to the diet at first by killing the small animals abundant on the savanna with bare hands, much as chimpanzees do. Much later there is evidence that large dead or dying animals were butchered with stone cutting tools.

Bipedalism, the ability to walk long distances while carrying infants, tools, gathered food and water, was essential to this new way of life, but to master these skills--upright walking, endurance, tool-using--required long child dependency which, in turn, necessitated sharing with young and group cooperation for their survival. The interaction of these factors resulted in natural selection for a larger brain to enhance communication among members in a community and to share information important for survival.

Early on, sharing became a fundamental basis for human social life. Food, initially shared regularly among mothers and their offspring, among females, expanded to sharing among male friends, and perhaps later still among members of a wider community. The movement of different groups through their home range, with transient mergings for mutual defense against predators, and sexual and social contacts, would have results in a rather complex social network. More like the pygmy chimpanzee pattern, these hominid groups included brothers and sons of older females and males who may have joined this core; the sociability of males in this evolving pattern may have made such males more desirable sexual partners to females.

The mastery of tool-using, initially of organic material or unmodified stones, improved their ability to obtain and prepare foods, and may have also improved their ability to defend themselves against predators. A shift in anticipatory behavior away from anatomy, together with increasing male sociability, probably accounts for the disappearance of the "fighting teeth," the long, sharp canines, that characterize male apes. Hominid males probably helped care for the young after weaning and shared food with their group on a regular basis, behaviors which are a departure from the ape pattern. Sexual relations were probably "casual." Pair bonds and a nuclear family would have been much too inflexible for a way of life involving constant movement and variable group size, and is inconsistent, too, with patterns observed among African apes and behaviors correlated with sexual dimorphism.

In the beginning, the hominid adaptation must have centered on gathering, rather than hunting, for the technological and social roots of hunting can be traced back to the gathering way of life. Many lines of evidence support this view: hominid's large grinding teeth; the omnivorous diet of closely related species, chimpanzees and contemporary gatherer-hunters; chimpanzees use of tools in obtaining and preparing food; and the absence of stone tools that would have been used to process or to hunt meat. Hunting eventually developed out of the coordination of a complex of human developments, already established for the earliest hominids: sharing, cooperation, incorporation of males into the social group, skill in making tools, using tools to obtain and prepare food, and a high level of communication about the environment. Hunting does not account for establishing the foundation for the hominid way of life.

The keynote of that early hominid adaptation was the ability to exploit available nutrients in a variety of ways. Versatility made possible the hominid expansion out of Africa throughout the tropical Old World and later into the northern and temperate regions. where food resources varied greatly: most plants in southeast Asia, mostly animals in the colder, temperate regions. The emergence of human culture, it can be argued, had its roots in that hominid gathering adaptation which emerged some 4 million years ago in East Africa.

Acknowledgments

I thank Catherine Borchert and Jerry Lowenstein for comments and discussion.

References

Behrensmeyer, A. K. 1976. Taphonomy and paleoecology in the hominid fossil record. Yearbook of Physical Anthropology 1975 19:36-50.

Bicchieri, M. G., ed. 1972. Hunters and Gatherers Today: A Socioeconomic Study of Eleven Such Cultures in the Twentieth Century. New York: Holt, Rinehart and Winston.

Binford, L. R. 1981. Bones. Ancient Men and Modern Myths. New York: Academic Press.

Boesch, C. and H. Boesch. 1981. Sex differences in the use of natural hammers by wild chimpanzees: A preliminary report. Journal of Human Evolution 10:585-593.

Bourliere, F. 1963. Observations on the ecology of some large African mammals. In: African Ecology and Human Evolution, ed., F. C. Howell and F. Bourliere, pp. 43-54. Chicago: Aldine.

Bourliere, F. 1973. The comparative ecology of rain forest mammals in Africa and tropical America: some introductory remarks. In: Tropical Forest Ecosystems in Africa and South America: A Comparative Review, ed., B. J. Meggers, E. S. Ayensu, and W. D. Duckworth, pp. 279-292. Washington, D.C.: Smithsonian Institution Press.

Bourliere, F. and M. Hadley. 1970. The ecology of tropical savannas. Annual Review of Ecology and Systematics 1:125-152.

Brain, C. K. 1970. New finds at the Swartkrans australopithecine site. Nature 225:1112-1119.

Brain, C. K. 1981. The Hunters or the Hunted? An Introduction to African Cave Taphonomy. Chicago: University Press.

Bunn, H. T. 1981. Archaeological evidence for meat-eating by Plio-Pleistocene hominids from Koobi Fora and Olduvai Gorge. Nature 291:574-577.

Butzer, K. W. 1977. Environment, culture and human evolution. American Scientist 65: 572-584.

Campbell, B. 1982. Humankind Emerging. Third Edition. Boston: Little Brown.

Coolidge, H. J., Jr. 1933. Pan paniscus: pigmy chimpanzee south of the Congo River. American Journal of Physical Anthropology 18:1-59.

Coppens, Y., F. C. Howell, G. L. Isaac, R.E.F. Leakey, eds. 1976. Earliest Man and Environments in the Lake Rudolf Basin. Stratigraphy, Paleoecology and Evolution. Chicago: University of Chicago Press.

Cramer, D. L. and A. L. Zihlman. 1978. Skeletal differences between pygmy (Pan paniscus) and common chimpanzees (Pan troglodytes). Folia primatologica 29(2):86-94.

Dahlberg, F. 1981. Woman the Gatherer. New Haven: Yale University Press.

Draper, P. and E. Cashdan. 1974. The impact of sedentism on !Kung socialization. Presented at the American Anthropological Association Meetings, 1974, Mexico City.

Gale, F., ed. 1974. Woman's Role in Aboriginal Society. 2nd edition. Australian Aboriginal Studies No. 36. Canberra: Australian Institute of Aboriginal Studies.

Goodall, J. van Lawick. 1967. Mother-offspring relationships in free-ranging chimpanzees. In: Primate Ethology, D. Morris, ed., pp. 287-346. London: Weidenfeld and Nicolson.

Goodall, J. van Lawick. 1968. The behaviour of free-living chimpanzees in the Gombe Stream Reserve. Animal Behaviour Monographs 1:165-311.

Goodall, J. and D. A. Hamburg. 1975. Chimpanzee behavior as a model for the behavior of early man. New evidence of possible origins of human behavior. American Handbook of Psychiatry 6:14-43.

Harding, R.S.O. and S. C. Strum. 1976. Predatory baboons of Kekopey. Natural History 85(3):46-53.

Harding, R.S.O. and G. Teleki. 1981. Omnivorous Primates. Gathering and Hunting in Human Evolution. New York: Columbia University Press.

Holloway, R. L. 1975. Early hominid endocasts: volumes, morphology and significance for hominid evolution. In: Primate Functional Morphology and Evolution, R. H. Tuttle, ed., pp. 393-415. The Hague: Mouton.

Howell, F. C. 1978. Hominidae. In: Evolution of African Mammals, V. J. Maglio & H.B.S. Cooke, eds., pp. 154-248. Cambridge, Mass.: Harvard University Press.

Howell, N. 1979. Demography of the Dobe !Kung. New York: Academic Press.

Isaac, G. L. 1978. The food-sharing behavior of protohuman hominids. Scientific American 238(4): 90-106.

Isaac, G. L. and D. C. Crader. 1981. To what extent were early hominids carnivorous? An archaeological perspective. In: Omnivorous Primates, R.S.O. Harding and G. Teleki, eds., pp. 37-103. New York: Columbia University Press.

Johanson, D. C. and M. Taieb. 1976. Plio-Pleistocene hominid discoveries in Hadar, Ethiopia. Nature 260:293-297.

Johanson, D. C. and T. D. White. 1979. A systematic assessment of early African hominids. Science 203:321-330.

Jolly, C. J., ed. 1978. Early Hominids of Africa. London: Duckworth.

Kano, T. 1980. Social behavior of wild pygmy chimpanzees (Pan paniscus) of Wamba: A preliminary report. Journal of Human Evolution 9:243-260.

Keeley, L. H. and N. Toth. 1981. Microwear polishes on early stone tools from Koobi Fora, Kenya. Nature 293:464-465.

Kimbel, W. H. and T. D. White. 1980. A reconstruction of the adult cranium of Australopithecus afarensis. American Journal of Physical Anthropology 52(2): 244. Abstract.

King. M.-C. and A. C. Wilson. 1975. Evolution at two levels in chimpanzees and humans. Science 188: 107-116.

Kitts, D. B. 1974. Paleontology and evolutionary theory. Evolution 28:458-472.

Kuroda, S. 1979. Grouping of the pygmy chimpanzees. Primates 20(2):161-183.

Kuroda, S. 1980. Social behavior of the pygmy chimpanzees. Primates 21(2):181-197.

Lancaster, J. B. 1968. On the evolution of tool-using behavior. American Anthropologist 70(1): 56-66.

Leakey, M. D. and R. L. Hay. 1979. Pliocene footprints in the Laetolil Beds at Laetoli, northern Tanzania. Nature 278:317-323.

Leakey, R.S.B. 1981. The Making of Mankind. New York: Dutton.

Lee, R. B. 1979. The !Kung San: Men, Women and Work in a Foraging Society. Cambridge: University Press.

Lee, R. B. and I. DeVore. 1976. Kalahari Hunter-Gatherers: Studies of the !Kung San and their Neighbors. Cambridge: Harvard University Press.

Le Gros Clark, W. E. 1967. Man-Apes or Ape-Men? The Story of Discoveries in Africa. New York: Holt, Rinehart and Winston.

Leutenegger, W. 1972. Newborn size of pelvic dimensions of Australopithecus. Nature 240:568-569.

Leutenegger, W. 1977. Sociobiological correlates of sexual dimorphism in body weight in South African Australopiths. South African Journal of Science 73:143-144.

Leutenegger, W. and J. T. Kelly. 1977. Relationship of sexual dimorphism in canine size and body size to social, behavioral and ecological correlates in anthropoid primates. Primates 18(1):117-136.

Linton, S. 1971. Woman the gatherer: male bias in anthropology. In: Women in Cross-Cultural Perspective, Sue-Ellen Jacobs, ed. Urbana: University of Illinois Press. Reprinted under Sally Slocum in Toward an Anthropology of Women,

R. R. Rapp, ed., pp. 36-50. New York: Monthly Review Press 1975.

Lovejoy, C. O. 1981. The origin of man. Science 211: 341-350.

Maglio, V. J. and H.S.B. Cooke. 1978. Evolution of African Mammals. Cambridge, Mass.: Harvard University Press.

Mann, A. 1975. Paleodemographic aspects of the South African australopithecines. Anthropology Publications No. 1. Philadelphia: University of Pennsylvania Press.

Marshall, L. 1976. The !Kung of Nyae Nyae. Cambridge, Mass.: Harvard University Press.

Marzke, M. 1982. Hominoid carpometacarpal joints, with special reference to the capitate and second and third metacarpals of Australopithecus afarensis. Manuscript.

McGrew, W. C. 1975. Patterns of plant food sharing by wild chimpanzees. Proceedings of the Vth International Congress, Primat. Soc., Nagoya, Japan, pp. 304-309. Basel: Karger.

McGrew, W. C. 1979. Evolutionary implications of sex difference in chimpanzee predation and tool use. In: The Great Apes, D. A. Hamburg and E. R. McCown, eds., pp. 440-463. Menlo Park, Calif.: Benjamin/Cummings.

Mellen, S.L.W. 1981. The Evolution of Love. San Francisco: W. H. Freeman.

Moss, C. 1975. Portraits in the Wild. Behavior Studies of East African Mammals. Boston: Houghton Mifflin.

Potts, R. and P. Shipman. 1981. Cutmarks made by stone tools on bones from Olduvai Gorge, Tanzania. Nature 291:577-580.

Reynolds, V. 1966. Open groups in hominid evolution. Man N.S. 1:441-452.

Robinson, J. T. 1972. Early Hominid Posture and Locomotion. Chicago: University of Chicago Press.

Sarich, V. M. and J. E. Cronin. 1976. Molecular systematics of the primates. In Molecular Anthropology, M. Goodman and R. E. Tashian, eds., pp. 141-170. New York: Plenum.

Shostak, M. 1981. Nisa. The life and words of a !Kung woman. Cambridge, Mass.: Harvard University Press.

Silberbauer, G. B. 1981. Hunter and Habitat in the Central Kalahari Desert. Cambridge: Cambridge University Press.

Silk, J. B. 1978. Patterns of food-sharing among mother and infant chimpanzees at Gombe National Park, Tanzania. Folia primatologica 29(2):129-141.

Tanner, N. and A. L. Zihlman. 1976. Women in evolution. Part I: Innovation and selection in human origins. Signs: Journal of Women in Culture and Society 1(3, part 1):585-608.

Taub, D. M. 1980. Female choice and mating strategies among wild barbary macaques (Macaca sylvanus L.). In The Macaques Studies in Ecology, Behavior, and Evolution, D. G. Lindburg, ed., pp. 287-344.

Teleki, G. 1973a. Group response to the accidental death of a chimpanzee in Gombe National Park, Tanzania. Folia primatologica 20(2-3):81-94.

Teleki, G. 1973b. The Predatory Behavior of Wild Chimpanzees. Lewisburg, Pa.: Bucknell University Press.

Teleki, G. 1974. Chimpanzee subsistence technology: materials and skills. Journal of Human Evolution 3:575-594.

Teleki, G. 1975. Primate subsistence patterns: collector-predators and gatherer-hunters. Journal of Human Evolution 4:125-184.

Tobias, P. V. 1975. Brain evolution in the hominoidea. In Primate Functional Morphology and Evolution, R. H. Tuttle, ed., pp. 353-392. The Hague: Mouton.

Tobias, P. V. 1980. "Australopithecus afarensis" and A. africanus: critique and an alternative hypothesis. Palaeont. afr. 23:1-17.

Trivers, R. L. 1972. Parental investment and sexual selection. In Sexual Selection and the Descent of Man 1871-1971, B. Campbell, ed., pp. 136-179. Chicago: Aldine-Atherton.

Tutin, C.E.G. 1979. Mating patterns and reproductive strategies in a community of wild chimpanzees (Pan troglodytes schweinfurthii). Behavioral Ecology and Sociobiology 6:29-38.

Wallace, J. A. 1978. Evolutionary trends in early hominid dentition. In Early Hominids of Africa, C. J. Jolly, ed., pp. 285-310. London: Duckworth.

Washburn, S. L. 1959. Speculations on the interrelations of the history of tools and biological evolution. Human Biology 31(1):21-31.

Washburn, S. L. 1960. Tools and human evolution. Scientific American 203(3)

Washburn, S. L. and DeVore. 1961. Social behavior of baboons and early man. In Social Life of Early Man, S. L. Washburn, ed. Chicago: Aldine.

Washburn, S. L. and S. Ranieri. 1981. Who brought home the bacon? New York Review of Books, September 24.

Wolpoff, M. H. 1973. Posterior tooth size, body size and diet in South African gracile australopithecines. American Journal of Physical Anthropology 39:375-394.

Wolpoff, M. H. 1975. Sexual dimorphism in the australopithecines. In Paleoanthropology, Morphology and Paleoecology, R. H. Tuttle, ed., pp. 245-284. The Hague: Mouton.

Wilson, P. 1980. Man, The Promising Primate. New Haven: Yale University Press.

Zihlman, A. L. 1976. Sexual dimorphism and its behavioral implications in early hominids. In Les Plus Anciens Hominides, P. V. Tobias and Y. Coppens, eds., pp. 268-293. Colloque VI, IX[e] Congres, Union Int. des Sci. prehist. et protohist. Paris: CNRS.

Zihlman, A. L. 1978. Motherhood in transition: from ape to human. In First Child and Family Formation, W. Miller and L. F. Newman, eds., pp. 35-50. Chapel Hill, N.C.: Carolina Population Center Publications.

Zihlman, A. L. 1979. Pygmy chimpanzees and early hominids. South African Journal of Science 75(4): 165-168.

Zihlman, A. L. 1981. Women as shapers of the human adaptation. In Woman the Gatherer, F. Dahlberg, ed., pp. 75-120. New Haven: Yale University Press.

Zihlman, A. L. 1982. The Human Evolution Coloring Book. New York: Harper and Row.

Zihlman, A. L. and L. Brunker. 1979. Hominid bipedalism: Then and now. Yearbook of Physical Anthropology 22:132-162.

Zihlman, A. L. and D. L. Cramer. 1978. A skeletal comparison between pygmy (Pan paniscus) and common chimpanzees (Pan troglodytes). Folia primatologica 29:86:94.

Zihlman, A. L., J. E. Cronin, D. L. Cramer, and V. M. Sarich. 1978. Pygmy chimpanzees as a possible prototype for the common ancestor of humans, chimpanzees and gorillas. Nature 275:744-746.